Student Edition

Eureka Math
Grade 8
Modules 1 & 2

Special thanks go to the Gordan A. Cain Center and to the Department of Mathematics at Louisiana State University for their support in the development of *Eureka Math*.

Published by Great Minds

Printed in the U.S.A.
This book may be purchased from the publisher at eureka-math.org
10 9 8 7 6 5 4 3 2 1

ISBN 978-1-63255-320-1

Lesson 1: Exponential Notation

Classwork

5^6 means $5 \times 5 \times 5 \times 5 \times 5 \times 5$, and $\left(\frac{9}{7}\right)^4$ means $\frac{9}{7} \times \frac{9}{7} \times \frac{9}{7} \times \frac{9}{7}$.

You have seen this kind of notation before; it is called exponential notation. In general, for any number x and any positive integer n,

$$x^n = \underbrace{(x \cdot x \cdots x)}_{n \text{ times}}.$$

The number x^n is called x raised to the nth power, where n is the exponent of x in x^n and x is the base of x^n.

Exercise 1

$\underbrace{4 \times \cdots \times 4}_{7 \text{ times}} =$

Exercise 2

$\underbrace{3.6 \times \cdots \times 3.6}_{\underline{\quad} \text{ times}} = 3.6^{47}$

Exercise 3

$\underbrace{(-11.63) \times \cdots \times (-11.63)}_{34 \text{ times}} =$

Exercise 4

$\underbrace{12 \times \cdots \times 12}_{\underline{\quad} \text{ times}} = 12^{15}$

Exercise 5

$\underbrace{(-5) \times \cdots \times (-5)}_{10 \text{ times}} =$

Exercise 6

$\underbrace{\frac{7}{2} \times \cdots \times \frac{7}{2}}_{21 \text{ times}} =$

Exercise 7

$\underbrace{(-13) \times \cdots \times (-13)}_{6 \text{ times}} =$

Exercise 8

$\underbrace{\left(-\frac{1}{14}\right) \times \cdots \times \left(-\frac{1}{14}\right)}_{10 \text{ times}} =$

Exercise 9

$\underbrace{x \cdot x \cdots x}_{185 \text{ times}} =$

Exercise 10

$\underbrace{x \cdot x \cdots x}_{\underline{\quad} \text{ times}} = x^n$

Lesson 1:   Exponential Notation

Exercise 11

Will these products be positive or negative? How do you know?

$$\underbrace{(-1) \times (-1) \times \cdots \times (-1)}_{12 \text{ times}} = (-1)^{12}$$

$$\underbrace{(-1) \times (-1) \times \cdots \times (-1)}_{13 \text{ times}} = (-1)^{13}$$

Exercise 12

Is it necessary to do all of the calculations to determine the sign of the product? Why or why not?

$$\underbrace{(-5) \times (-5) \times \cdots \times (-5)}_{95 \text{ times}} = (-5)^{95}$$

$$\underbrace{(-1.8) \times (-1.8) \times \cdots \times (-1.8)}_{122 \text{ times}} = (-1.8)^{122}$$

Exercise 13

Fill in the blanks indicating whether the number is positive or negative.

If n is a positive even number, then $(-55)^n$ is _____.

If n is a positive odd number, then $(-72.4)^n$ is _____.

Exercise 14

Josie says that $\underbrace{(-15) \times \cdots \times (-15)}_{6\ times} = -15^6$. Is she correct? How do you know?

Problem Set

1. Use what you know about exponential notation to complete the expressions below.

$$\underbrace{(-5) \times \cdots \times (-5)}_{17\ times} =$$
$$\underbrace{3.7 \times \cdots \times 3.7}_{\_\_\_\ times} = 3.7^{19}$$

$$\underbrace{7 \times \cdots \times 7}_{\_\_\_\ times} = 7^{45}$$
$$\underbrace{6 \times \cdots \times 6}_{4\ times} =$$

$$\underbrace{4.3 \times \cdots \times 4.3}_{13\ times} =$$
$$\underbrace{(-1.1) \times \cdots \times (-1.1)}_{9\ times} =$$

$$\underbrace{\left(\tfrac{2}{3}\right) \times \cdots \times \left(\tfrac{2}{3}\right)}_{19\ times} =$$
$$\underbrace{\left(-\tfrac{11}{5}\right) \times \cdots \times \left(-\tfrac{11}{5}\right)}_{\_\_\_\ times} = \left(-\tfrac{11}{5}\right)^{x}$$

$$\underbrace{(-12) \times \cdots \times (-12)}_{\_\_\_\ times} = (-12)^{15}$$
$$\underbrace{a \times \cdots \times a}_{m\ times} =$$

2. Write an expression with (-1) as its base that will produce a positive product, and explain why your answer is valid.

3. Write an expression with (-1) as its base that will produce a negative product, and explain why your answer is valid.

4. Rewrite each number in exponential notation using 2 as the base.

    $8 =$ $16 =$ $32 =$

    $64 =$ $128 =$ $256 =$

5. Tim wrote 16 as $(-2)^4$. Is he correct? Explain.

6. Could -2 be used as a base to rewrite 32? 64? Why or why not?

Lesson 2: Multiplication of Numbers in Exponential Form

Classwork

In general, if x is any number and m, n are positive integers, then

$$x^m \cdot x^n = x^{m+n}$$

because

$$x^m \times x^n = \underbrace{(x \cdots x)}_{m \text{ times}} \times \underbrace{(x \cdots x)}_{n \text{ times}} = \underbrace{(x \cdots x)}_{m+n \text{ times}} = x^{m+n}.$$

Exercise 1

$14^{23} \times 14^8 =$

Exercise 2

$(-72)^{10} \times (-72)^{13} =$

Exercise 3

$5^{94} \times 5^{78} =$

Exercise 4

$(-3)^9 \times (-3)^5 =$

Exercise 5

Let a be a number.

$a^{23} \cdot a^8 =$

Exercise 6

Let f be a number.

$f^{10} \cdot f^{13} =$

Exercise 7

Let b be a number.

$b^{94} \cdot b^{78} =$

Exercise 8

Let x be a positive integer. If $(-3)^9 \times (-3)^x = (-3)^{14}$, what is x?

What would happen if there were more terms with the same base? Write an equivalent expression for each problem.

Exercise 9

$9^4 \times 9^6 \times 9^{13} =$

Exercise 10

$2^3 \times 2^5 \times 2^7 \times 2^9 =$

Can the following expressions be written in simpler form? If so, write an equivalent expression. If not, explain why not.

Exercise 11

$6^5 \times 4^9 \times 4^3 \times 6^{14} =$

Exercise 14

$2^4 \times 8^2 = 2^4 \times 2^6 =$

Exercise 12

$(-4)^2 \cdot 17^5 \cdot (-4)^3 \cdot 17^7 =$

Exercise 15

$3^7 \times 9 = 3^7 \times 3^2 =$

Exercise 13

$15^2 \cdot 7^2 \cdot 15 \cdot 7^4 =$

Exercise 16

$5^4 \times 2^{11} =$

Exercise 17

Let x be a number. Rewrite the expression in a simpler form.

$(2x^3)(17x^7) =$

Exercise 18

Let a and b be numbers. Use the distributive law to rewrite the expression in a simpler form.

$a(a + b) =$

Exercise 19

Let a and b be numbers. Use the distributive law to rewrite the expression in a simpler form.

$b(a + b) =$

Exercise 20

Let a and b be numbers. Use the distributive law to rewrite the expression in a simpler form.

$(a + b)(a + b) =$

In general, if x is nonzero and m, n are positive integers, then

$$\frac{x^m}{x^n} = x^{m-n}.$$

Exercise 21

$\dfrac{7^9}{7^6} =$

Exercise 23

$\dfrac{\left(\frac{8}{5}\right)^9}{\left(\frac{8}{5}\right)^2} =$

Exercise 22

$\dfrac{(-5)^{16}}{(-5)^7} =$

Exercise 24

$\dfrac{13^5}{13^4} =$

Exercise 25

Let a, b be nonzero numbers. What is the following number?

$$\frac{\left(\frac{a}{b}\right)^9}{\left(\frac{a}{b}\right)^2} =$$

Exercise 26

Let x be a nonzero number. What is the following number?

$$\frac{x^5}{x^4} =$$

Can the following expressions be written in simpler forms? If yes, write an equivalent expression for each problem. If not, explain why not.

Exercise 27

$$\frac{2^7}{4^2} = \frac{2^7}{2^4} =$$

Exercise 28

$$\frac{3^{23}}{27} = \frac{3^{23}}{3^3} =$$

Exercise 29

$$\frac{3^5 \cdot 2^8}{3^2 \cdot 2^3} =$$

Exercise 30

$$\frac{(-2)^7 \cdot 95^5}{(-2)^5 \cdot 95^4} =$$

Exercise 31

Let x be a number. Write each expression in a simpler form.

a. $\dfrac{5}{x^3}(3x^8) =$

b. $\dfrac{5}{x^3}(-4x^6) =$

c. $\dfrac{5}{x^3}(11x^4) =$

Exercise 32

Anne used an online calculator to multiply $2\,000\,000\,000 \times 2\,000\,000\,000\,000$. The answer showed up on the calculator as $4e+21$, as shown below. Is the answer on the calculator correct? How do you know?

2000000000 × 2000000000000 =

4e+21

Rad		x!	(	)	%	AC
Inv	sin	ln	7	8	9	÷
π	cos	log	4	5	6	×
e	tan	√	1	2	3	−
Ans	EXP	x^y	0	.	=	+

EUREKA
MATH™

Problem Set

1. A certain ball is dropped from a height of x feet. It always bounces up to $\frac{2}{3}x$ feet. Suppose the ball is dropped from 10 feet and is stopped exactly when it touches the ground after the 30th bounce. What is the total distance traveled by the ball? Express your answer in exponential notation.

Bounce	Computation of Distance Traveled in Previous Bounce	Total Distance Traveled (in feet)
1		
2		
3		
4		
30		
n		

2. If the same ball is dropped from 10 feet and is caught exactly at the highest point after the 25th bounce, what is the total distance traveled by the ball? Use what you learned from the last problem.

3. Let a and b be numbers and $b \neq 0$, and let m and n be positive integers. Write each expression using the fewest number of bases possible:

$(-19)^5 \cdot (-19)^{11} =$	$2.7^5 \times 2.7^3 =$
$\dfrac{7^{10}}{7^3} =$	$\left(\dfrac{1}{5}\right)^2 \cdot \left(\dfrac{1}{5}\right)^{15} =$
$\left(-\dfrac{9}{7}\right)^m \cdot \left(-\dfrac{9}{7}\right)^n =$	$\dfrac{ab^3}{b^2} =$

4. Let the dimensions of a rectangle be $(4 \times (871209)^5 + 3 \times 49762105)$ ft. by $(7 \times (871209)^3 - (49762105)^4)$ ft. Determine the area of the rectangle. (Hint: You do not need to expand all the powers.)

5. A rectangular area of land is being sold off in smaller pieces. The total area of the land is 2^{15} square miles. The pieces being sold are 8^3 square miles in size. How many smaller pieces of land can be sold at the stated size? Compute the actual number of pieces.

Lesson 3: Numbers in Exponential Form Raised to a Power

Classwork

For any number x and any positive integers m and n,

$$(x^m)^n = x^{nm}$$

because

$$(x^m)^n = \underbrace{(x \cdot x \cdots x)}_{m \text{ times}}{}^n$$

$$= \underbrace{\underbrace{(x \cdot x \cdots x)}_{m \text{ times}} \times \cdots \times \underbrace{(x \cdot x \cdots x)}_{m \text{ times}}}_{n \text{ times}}$$

$$= x^{nm}.$$

Exercise 1

$(15^3)^9 =$

Exercise 3

$(3.4^{17})^4 =$

Exercise 2

$\left((-2)^5\right)^8 =$

Exercise 4

Let s be a number.

$(s^{17})^4 =$

Exercise 5

Sarah wrote $(3^5)^7 = 3^{12}$. Correct her mistake. Write an exponential expression using a base of 3 and exponents of 5, 7, and 12 that would make her answer correct.

Exercise 6

A number y satisfies $y^{24} - 256 = 0$. What equation does the number $x = y^4$ satisfy?

EUREKA
MATH™

Lesson 3:   Numbers in Exponential Form Raised to a Power

For any numbers x and y, and positive integer n,

$$(xy)^n = x^n y^n$$

because

$$(xy)^n = \underbrace{(xy) \cdots (xy)}_{n \text{ times}}$$
$$= \underbrace{(x \cdot x \cdots x)}_{n \text{ times}} \cdot \underbrace{(y \cdot y \cdots y)}_{n \text{ times}}$$
$$= x^n y^n.$$

Exercise 7

$(11 \times 4)^9 =$

Exercise 8

$(3^2 \times 7^4)^5 =$

Exercise 9

Let a, b, and c be numbers.

$(3^2 a^4)^5 =$

Exercise 10

Let x be a number.

$(5x)^7 =$

Exercise 11

Let x and y be numbers.

$(5xy^2)^7 =$

Exercise 12

Let a, b, and c be numbers.

$(a^2 b c^3)^4 =$

Exercise 13

Let x and y be numbers, $y \neq 0$, and let n be a positive integer. How is $\left(\dfrac{x}{y}\right)^n$ related to x^n and y^n?

Problem Set

1. Show (prove) in detail why $(2 \cdot 3 \cdot 7)^4 = 2^4 3^4 7^4$.

2. Show (prove) in detail why $(xyz)^4 = x^4 y^4 z^4$ for any numbers x, y, z.

3. Show (prove) in detail why $(xyz)^n = x^n y^n z^n$ for any numbers $x, y,$ and z and for any positive integer n.

Lesson 4: Numbers Raised to the Zeroth Power

Classwork

We have shown that for any numbers x, y, and any positive integers m, n, the following holds

$$x^m \cdot x^n = x^{m+n} \tag{1}$$

$$(x^m)^n = x^{mn} \tag{2}$$

$$(xy)^n = x^n y^n. \tag{3}$$

Definition: _____

Exercise 1

List all possible cases of whole numbers m and n for identity (1). More precisely, when $m > 0$ and $n > 0$, we already know that (1) is correct. What are the other possible cases of m and n for which (1) is yet to be verified?

Exercise 2

Check that equation (1) is correct for each of the cases listed in Exercise 1.

Exercise 3

Do the same with equation (2) by checking it case-by-case.

Exercise 4

Do the same with equation (3) by checking it case-by-case.

Exercise 5

Write the expanded form of 8,374 using exponential notation.

Exercise 6

Write the expanded form of 6,985,062 using exponential notation.

Problem Set

Let x, y be numbers $(x, y \neq 0)$. Simplify each of the following expressions.

1. $$\frac{y^{12}}{y^{12}} =$$	2. $$9^{15} \cdot \frac{1}{9^{15}} =$$
3. $$(7(123456.789)^4)^0 =$$	4. $$2^2 \cdot \frac{1}{2^5} \cdot 2^5 \cdot \frac{1}{2^2} = \frac{2^2}{2^2} \cdot \frac{2^5}{2^5}$$

5.
$$\frac{x^{41}}{y^{15}} \cdot \frac{y^{15}}{x^{41}} = \frac{x^{41} \cdot y^{15}}{y^{15} \cdot x^{41}}$$

Number Correct: _____

Applying Properties of Exponents to Generate Equivalent Expressions—Round 1

Directions: Simplify each expression using the laws of exponents. Use the least number of bases possible and only positive exponents. All letters denote numbers.

1.	$2^2 \cdot 2^3 =$		23.	$6^3 \cdot 6^2 =$	
2.	$2^2 \cdot 2^4 =$		24.	$6^2 \cdot 6^3 =$	
3.	$2^2 \cdot 2^5 =$		25.	$(-8)^3 \cdot (-8)^7 =$	
4.	$3^7 \cdot 3^1 =$		26.	$(-8)^7 \cdot (-8)^3 =$	
5.	$3^8 \cdot 3^1 =$		27.	$(0.2)^3 \cdot (0.2)^7 =$	
6.	$3^9 \cdot 3^1 =$		28.	$(0.2)^7 \cdot (0.2)^3 =$	
7.	$7^6 \cdot 7^2 =$		29.	$(-2)^{12} \cdot (-2)^1 =$	
8.	$7^6 \cdot 7^3 =$		30.	$(-2.7)^{12} \cdot (-2.7)^1 =$	
9.	$7^6 \cdot 7^4 =$		31.	$1.1^6 \cdot 1.1^9 =$	
10.	$11^{15} \cdot 11 =$		32.	$57^6 \cdot 57^9 =$	
11.	$11^{16} \cdot 11 =$		33.	$x^6 \cdot x^9 =$	
12.	$2^{12} \cdot 2^2 =$		34.	$2^7 \cdot 4 =$	
13.	$2^{12} \cdot 2^4 =$		35.	$2^7 \cdot 4^2 =$	
14.	$2^{12} \cdot 2^6 =$		36.	$2^7 \cdot 16 =$	
15.	$99^5 \cdot 99^2 =$		37.	$16 \cdot 4^3 =$	
16.	$99^6 \cdot 99^3 =$		38.	$3^2 \cdot 9 =$	
17.	$99^7 \cdot 99^4 =$		39.	$3^2 \cdot 27 =$	
18.	$5^8 \cdot 5^2 =$		40.	$3^2 \cdot 81 =$	
19.	$6^8 \cdot 6^2 =$		41.	$5^4 \cdot 25 =$	
20.	$7^8 \cdot 7^2 =$		42.	$5^4 \cdot 125 =$	
21.	$r^8 \cdot r^2 =$		43.	$8 \cdot 2^9 =$	
22.	$s^8 \cdot s^2 =$		44.	$16 \cdot 2^9 =$	

Number Correct: _____

Directions. Simplify each expression using the laws of exponents. Use the least number of bases possible and only positive exponents. All letters denote positive numbers.

Number Correct: _____

Improvement: _____

Applying Properties of Exponents to Generate Equivalent Expressions—Round 2

Directions: Simplify each expression using the laws of exponents. Use the least number of bases possible and only positive exponents. All letters denote numbers.

1.	$5^2 \cdot 5^3 =$		23.	$7^3 \cdot 7^2 =$	
2.	$5^2 \cdot 5^4 =$		24.	$7^2 \cdot 7^3 =$	
3.	$5^2 \cdot 5^5 =$		25.	$(-4)^3 \cdot (-4)^{11} =$	
4.	$2^7 \cdot 2^1 =$		26.	$(-4)^{11} \cdot (-4)^3 =$	
5.	$2^8 \cdot 2^1 =$		27.	$(0.2)^3 \cdot (0.2)^{11} =$	
6.	$2^9 \cdot 2^1 =$		28.	$(0.2)^{11} \cdot (0.2)^3 =$	
7.	$3^6 \cdot 3^2 =$		29.	$(-2)^9 \cdot (-2)^5 =$	
8.	$3^6 \cdot 3^3 =$		30.	$(-2.7)^5 \cdot (-2.7)^9 =$	
9.	$3^6 \cdot 3^4 =$		31.	$3.1^6 \cdot 3.1^6 =$	
10.	$7^{15} \cdot 7 =$		32.	$57^6 \cdot 57^6 =$	
11.	$7^{16} \cdot 7 =$		33.	$z^6 \cdot z^6 =$	
12.	$11^{12} \cdot 11^2 =$		34.	$4 \cdot 2^9 =$	
13.	$11^{12} \cdot 11^4 =$		35.	$4^2 \cdot 2^9 =$	
14.	$11^{12} \cdot 11^6 =$		36.	$16 \cdot 2^9 =$	
15.	$23^5 \cdot 23^2 =$		37.	$16 \cdot 4^3 =$	
16.	$23^6 \cdot 23^3 =$		38.	$9 \cdot 3^5 =$	
17.	$23^7 \cdot 23^4 =$		39.	$3^5 \cdot 9 =$	
18.	$13^7 \cdot 13^3 =$		40.	$3^5 \cdot 27 =$	
19.	$15^7 \cdot 15^3 =$		41.	$5^7 \cdot 25 =$	
20.	$17^7 \cdot 17^3 =$		42.	$5^7 \cdot 125 =$	
21.	$x^7 \cdot x^3 =$		43.	$2^{11} \cdot 4 =$	
22.	$y^7 \cdot y^3 =$		44.	$2^{11} \cdot 16 =$	

Lesson 5: Negative Exponents and the Laws of Exponents

Classwork

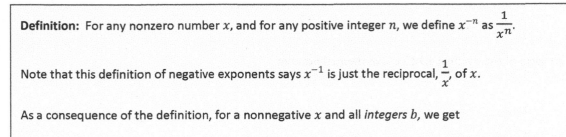

Definition: For any nonzero number x, and for any positive integer n, we define x^{-n} as $\frac{1}{x^n}$.

Note that this definition of negative exponents says x^{-1} is just the reciprocal, $\frac{1}{x}$, of x.

As a consequence of the definition, for a nonnegative x and all *integers b*, we get

$$x^{-b} = \frac{1}{x^b}$$

Exercise 1

Verify the general statement $x^{-b} = \frac{1}{x^b}$ for $x = 3$ and $b = -5$.

Exercise 2

What is the value of (3×10^{-2})?

Exercise 3

What is the value of (3×10^{-5})?

Exercise 4

Write the complete expanded form of the decimal 4.728 in exponential notation.

For Exercises 5–10, write an equivalent expression, in exponential notation, to the one given, and simplify as much as possible.

Exercise 5

$5^{-3} =$

Exercise 6

$\dfrac{1}{8^9} =$

Exercise 7

$3 \cdot 2^{-4} =$

Exercise 8

Let x be a nonzero number.

$x^{-3} =$

Exercise 9

Let x be a nonzero number.

$\dfrac{1}{x^9} =$

Exercise 10

Let x, y be two nonzero numbers.

$xy^{-4} =$

We accept that for nonzero numbers x and y and all integers a and b,

$$x^a \cdot x^b = x^{a+b}$$

$$(x^b)^a = x^{ab}$$

$$(xy)^a = x^a y^a.$$

We claim

$$\frac{x^a}{x^b} = x^{a-b} \qquad\qquad \text{for all integers } a, b.$$

$$\left(\frac{x}{y}\right)^a = \frac{x^a}{y^a} \qquad\qquad \text{for any integer } a.$$

Exercise 11

$$\frac{19^2}{19^5} =$$

Exercise 12

$$\frac{17^{16}}{17^{-3}} =$$

Exercise 13

If we let $b = -1$ in (11), a be any integer, and y be any nonzero number, what do we get?

Exercise 14

Show directly that $\left(\frac{7}{5}\right)^{-4} = \frac{7^{-4}}{5^{-4}}$.

Problem Set

1. Compute: $3^3 \times 3^2 \times 3^1 \times 3^0 \times 3^{-1} \times 3^{-2} =$

    Compute: $5^2 \times 5^{10} \times 5^8 \times 5^0 \times 5^{-10} \times 5^{-8} =$

    Compute for a nonzero number, a: $a^m \times a^n \times a^l \times a^{-n} \times a^{-m} \times a^{-l} \times a^0 =$

2. Without using (10), show directly that $(17.6^{-1})^8 = 17.6^{-8}$.

3. Without using (10), show (prove) that for any whole number n and any positive number y, $(y^{-1})^n = y^{-n}$.

4. Without using (13), show directly without using (13) that $\dfrac{2.8^{-5}}{2.8^7} = 2.8^{-12}$.

Equation Reference Sheet

For any numbers x, y [$x \neq 0$ in (4) and $y \neq 0$ in (5)] and any positive integers m, n, the following holds:

$$x^m \cdot x^n = x^{m+n} \tag{1}$$

$$(x^m)^n = x^{mn} \tag{2}$$

$$(xy)^n = x^n y^n \tag{3}$$

$$\frac{x^m}{x^n} = x^{m-n} \tag{4}$$

$$\left(\frac{x}{y}\right)^n = \frac{x^n}{y^n} \tag{5}$$

For any numbers x, y and for all whole numbers m, n, the following holds:

$$x^m \cdot x^n = x^{m+n} \tag{6}$$

$$(x^m)^n = x^{mn} \tag{7}$$

$$(xy)^n = x^n y^n \tag{8}$$

For any nonzero number x and all integers b, the following holds:

$$x^{-b} = \frac{1}{x^b} \tag{9}$$

For any numbers x, y and all integers a, b, the following holds:

$$x^a \cdot x^b = x^{a+b} \tag{10}$$

$$(x^b)^a = x^{ab} \tag{11}$$

$$(xy)^a = x^a y^a \tag{12}$$

$$\frac{x^a}{x^b} = x^{a-b} \qquad x \neq 0 \tag{13}$$

$$\left(\frac{x}{y}\right)^a = \frac{x^a}{y^a} \qquad x, y \neq 0 \tag{14}$$

Lesson 6: Proofs of Laws of Exponents

Classwork

The Laws of Exponents

For $x, y \neq 0$, and all integers a, b, the following holds:

$$x^a \cdot x^b = x^{a+b}$$

$$(x^b)^a = x^{ab}$$

$$(xy)^a = x^a y^a.$$

Facts we will use to prove (11):

(A) (11) is already known to be true when the integers a and b satisfy $a \geq 0, b \geq 0$.

(B) $x^{-m} = \dfrac{1}{x^m}$ for any whole number m.

(C) $\left(\dfrac{1}{x}\right)^m = \dfrac{1}{x^m}$ for any whole number m.

Exercise 1

Show that **(C)** is implied by equation (5) of Lesson 4 when $m > 0$, and explain why **(C)** continues to hold even when $m = 0$.

Exercise 2

Show that **(B)** is in fact a special case of (11) by rewriting it as $(x^m)^{-1} = x^{(-1)m}$ for any whole number m, so that if $b = m$ (where m is a whole number) and $a = -1$, (11) becomes **(B)**.

Exercise 3

Show that **(C)** is a special case of (11) by rewriting **(C)** as $(x^{-1})^m = x^{m(-1)}$ for any whole number m. Thus, **(C)** is the special case of (11) when $b = -1$ and $a = m$, where m is a whole number.

Exercise 4

Proof of Case (iii): Show that when $a < 0$ and $b \geq 0$, $(x^b)^a = x^{ab}$ is still valid. Let $a = -c$ for some positive integer c. Show that the left and right sides of $(x^b)^a = x^{ab}$ are equal.

Problem Set

1. You sent a photo of you and your family on vacation to seven Facebook friends. If each of them sends it to five of their friends, and each of those friends sends it to five of their friends, and those friends send it to five more, how many people (not counting yourself) will see your photo? No friend received the photo twice. Express your answer in exponential notation.

# of New People to View Your Photo	Total # of People to View Your Photo

2. Show directly, without using (11), that $(1.27^{-36})^{85} = 1.27^{-36 \cdot 85}$.

3. Show directly that $\left(\frac{2}{13}\right)^{-127} \cdot \left(\frac{2}{13}\right)^{-56} = \left(\frac{2}{13}\right)^{-183}$.

4. Prove for any nonzero number x, $x^{-127} \cdot x^{-56} = x^{-183}$.

5. Prove for any nonzero number x, $x^{-m} \cdot x^{-n} = x^{-m-n}$ for positive integers m and n.

6. Which of the preceding four problems did you find easiest to do? Explain.

7. Use the properties of exponents to write an equivalent expression that is a product of distinct primes, each raised to an integer power.

$$\frac{10^5 \cdot 9^2}{6^4} =$$

Lesson 7: Magnitude

Classwork

Fact 1: *The number 10^n, for arbitrarily large positive integers n, is a* big number in the sense that given a number M (no matter how big it is) there is a power of 10 that exceeds M.

Fact 2: *The number 10^{-n}, for arbitrarily large positive integers n, is a* small number in the sense that given a positive number S (no matter how small it is), there is a (negative) power of 10 that is smaller than S.

Exercise 1

Let $M = 993,456,789,098,765$. Find the smallest power of 10 that will exceed M.

Exercise 2

Let $M = 78,491 \dfrac{899}{987}$. Find the smallest power of 10 that will exceed M.

Exercise 3

Let M be a positive integer. Explain how to find the smallest power of 10 that exceeds it.

Exercise 4

The chance of you having the same DNA as another person (other than an identical twin) is approximately 1 in 10 trillion (one trillion is a 1 followed by 12 zeros). Given the fraction, express this very small number using a negative power of 10.

$$\frac{1}{10\,000\,000\,000\,000}$$

Exercise 5

The chance of winning a big lottery prize is about 10^{-8}, and the chance of being struck by lightning in the U.S. in any given year is about $0.000\,001$. Which do you have a greater chance of experiencing? Explain.

Exercise 6

There are about 100 million smartphones in the U.S. Your teacher has one smartphone. What share of U.S. smartphones does your teacher have? Express your answer using a negative power of 10.

Problem Set

1. What is the smallest power of 10 that would exceed 987,654,321,098,765,432?

2. What is the smallest power of 10 that would exceed 999,999,999,991?

3. Which number is equivalent to 0.000 000 1: 10^7 or 10^{-7}? How do you know?

4. Sarah said that 0.000 01 is bigger than 0.001 because the first number has more digits to the right of the decimal point. Is Sarah correct? Explain your thinking using negative powers of 10 and the number line.

5. Order the following numbers from least to greatest:

    $$10^5 \qquad 10^{-99} \qquad 10^{-17} \qquad 10^{14} \qquad 10^{-5} \qquad 10^{30}$$

Lesson 8: Estimating Quantities

Exercise 1

The Federal Reserve states that the average household in January of 2013 had $7,122 in credit card debt. About how many times greater is the U.S. national debt, which is $16,755,133,009,522? Rewrite each number to the nearest power of 10 that exceeds it, and then compare.

Exercise 2

There are about 3,000,000 students attending school, kindergarten through Grade 12, in New York. Express the number of students as a single-digit integer times a power of 10.

The average number of students attending a middle school in New York is 8×10^2. How many times greater is the overall number of K–12 students compared to the average number of middle school students?

Exercise 3

A conservative estimate of the number of stars in the universe is 6×10^{22}. The average human can see about 3,000 stars at night with his naked eye. About how many times more stars are there in the universe compared to the stars a human can actually see?

Exercise 4

The estimated world population in 2011 was 7×10^9. Of the total population, 682 million of those people were left-handed. Approximately what percentage of the world population is left-handed according to the 2011 estimation?

Exercise 5

The average person takes about 30,000 breaths per day. Express this number as a single-digit integer times a power of 10.

If the average American lives about 80 years (or about 30,000 days), how many total breaths will a person take in her lifetime?

Problem Set

1. The Atlantic Ocean region contains approximately 2×10^{16} gallons of water. Lake Ontario has approximately 8,000,000,000 gallons of water. How many Lake Ontarios would it take to fill the Atlantic Ocean region in terms of gallons of water?

2. U.S. national forests cover approximately 300,000 square miles. Conservationists want the total square footage of forests to be $300,000^2$ square miles. When Ivanna used her phone to do the calculation, her screen showed the following:

    a. What does the answer on her screen mean? Explain how you know.

    b. Given that the U.S. has approximately 4 million square miles of land, is this a reasonable goal for conservationists? Explain.

3. The average American is responsible for about 20,000 kilograms of carbon emission pollution each year. Express this number as a single-digit integer times a power of 10.

4. The United Kingdom is responsible for about 1×10^4 kilograms of carbon emission pollution each year. Which country is responsible for greater carbon emission pollution each year? By how much?

Number Correct: _____

Applying Properties of Exponents to Generate Equivalent Expressions—Round 1

Directions: Simplify each expression using the laws of exponents. Use the least number of bases possible and only positive exponents. When appropriate, express answers without parentheses or as equal to 1. All letters denote numbers.

1.	$4^5 \cdot 4^{-4} =$		23.	$\left(\frac{1}{2}\right)^6 =$	
2.	$4^5 \cdot 4^{-3} =$		24.	$(3x)^5 =$	
3.	$4^5 \cdot 4^{-2} =$		25.	$(3x)^7 =$	
4.	$7^{-4} \cdot 7^{11} =$		26.	$(3x)^9 =$	
5.	$7^{-4} \cdot 7^{10} =$		27.	$(8^{-2})^3 =$	
6.	$7^{-4} \cdot 7^9 =$		28.	$(8^{-3})^3 =$	
7.	$9^{-4} \cdot 9^{-3} =$		29.	$(8^{-4})^3 =$	
8.	$9^{-4} \cdot 9^{-2} =$		30.	$(22^0)^{50} =$	
9.	$9^{-4} \cdot 9^{-1} =$		31.	$(22^0)^{55} =$	
10.	$9^{-4} \cdot 9^0 =$		32.	$(22^0)^{60} =$	
11.	$5^0 \cdot 5^1 =$		33.	$\left(\frac{1}{11}\right)^{-5} =$	
12.	$5^0 \cdot 5^2 =$		34.	$\left(\frac{1}{11}\right)^{-6} =$	
13.	$5^0 \cdot 5^3 =$		35.	$\left(\frac{1}{11}\right)^{-7} =$	
14.	$(12^3)^9 =$		36.	$\frac{56^{-23}}{56^{-34}} =$	
15.	$(12^3)^{10} =$		37.	$\frac{87^{-12}}{87^{-34}} =$	
16.	$(12^3)^{11} =$		38.	$\frac{23^{-15}}{23^{-17}} =$	
17.	$(7^{-3})^{-8} =$		39.	$(-2)^{-12} \cdot (-2)^1 =$	
18.	$(7^{-3})^{-9} =$		40.	$\frac{2y}{y^3} =$	
19.	$(7^{-3})^{-10} =$		41.	$\frac{5xy^7}{15x^7y} =$	
20.	$\left(\frac{1}{2}\right)^9 =$		42.	$\frac{16x^6y^9}{8x^{-5}y^{-11}} =$	
21.	$\left(\frac{1}{2}\right)^8 =$		43.	$(2^3 \cdot 4)^{-5} =$	
22.	$\left(\frac{1}{2}\right)^7 =$		44.	$(9^{-8})(27^{-2}) =$	

Number Correct: _____

Improvement: _____

Applying Properties of Exponents to Generate Equivalent Expressions—Round 2

Directions: Simplify each expression using the laws of exponents. Use the least number of bases possible and only positive exponents. When appropriate, express answers without parentheses or as equal to 1. All letters denote numbers.

1.	$11^5 \cdot 11^{-4} =$		23.	$\left(\frac{3}{7}\right)^5 =$	
2.	$11^5 \cdot 11^{-3} =$		24.	$(18xy)^5 =$	
3.	$11^5 \cdot 11^{-2} =$		25.	$(18xy)^7 =$	
4.	$7^{-7} \cdot 7^9 =$		26.	$(18xy)^9 =$	
5.	$7^{-8} \cdot 7^9 =$		27.	$(5.2^{-2})^3 =$	
6.	$7^{-9} \cdot 7^9 =$		28.	$(5.2^{-3})^3 =$	
7.	$(-6)^{-4} \cdot (-6)^{-3} =$		29.	$(5.2^{-4})^3 =$	
8.	$(-6)^{-4} \cdot (-6)^{-2} =$		30.	$(22^6)^0 =$	
9.	$(-6)^{-4} \cdot (-6)^{-1} =$		31.	$(22^{12})^0 =$	
10.	$(-6)^{-4} \cdot (-6)^0 =$		32.	$(22^{18})^0 =$	
11.	$x^0 \cdot x^1 =$		33.	$\left(\frac{4}{5}\right)^{-5} =$	
12.	$x^0 \cdot x^2 =$		34.	$\left(\frac{4}{5}\right)^{-6} =$	
13.	$x^0 \cdot x^3 =$		35.	$\left(\frac{4}{5}\right)^{-7} =$	
14.	$(12^5)^9 =$		36.	$\left(\frac{6^{-2}}{7^5}\right)^{-11} =$	
15.	$(12^6)^9 =$		37.	$\left(\frac{6^{-2}}{7^5}\right)^{-12} =$	
16.	$(12^7)^9 =$		38.	$\left(\frac{6^{-2}}{7^5}\right)^{-13} =$	
17.	$(7^{-3})^{-4} =$		39.	$\left(\frac{6^{-2}}{7^5}\right)^{-15} =$	
18.	$(7^{-4})^{-4} =$		40.	$\frac{42ab^{10}}{14a^{-9}b} =$	
19.	$(7^{-5})^{-4} =$		41.	$\frac{5xy^7}{25x^7y} =$	
20.	$\left(\frac{3}{7}\right)^8 =$		42.	$\frac{22a^{15}b^{32}}{121ab^{-5}} =$	
21.	$\left(\frac{3}{7}\right)^7 =$		43.	$(7^{-8} \cdot 49)^{-5} =$	
22.	$\left(\frac{3}{7}\right)^6 =$		44.	$(36^9)(216^{-2}) =$	

Lesson 9: Scientific Notation

Classwork

> A positive, finite decimal s is said to be written in scientific notation if it is expressed as a product $d \times 10^n$, where d is a finite decimal so that $1 \leq d < 10$, and n is an integer.
>
> The integer n is called the order of magnitude of the decimal $d \times 10^n$.

Are the following numbers written in scientific notation? If not, state the reason.

Exercise 1

1.908×10^{17}

Exercise 4

$4.0701 + 10^7$

Exercise 2

0.325×10^{-2}

Exercise 5

18.432×5^8

Exercise 3

7.99×10^{32}

Exercise 6

8×10^{-11}

Use the table below to complete Exercises 7 and 8.

The table below shows the debt of the three most populous states and the three least populous states.

State	Debt (in dollars)	Population (2012)
California	407,000,000,000	38,000,000
New York	337,000,000,000	19,000,000
Texas	276,000,000,000	26,000,000
North Dakota	4,000,000,000	690,000
Vermont	4,000,000,000	626,000
Wyoming	2,000,000,000	576,000

Exercise 7

a. What is the sum of the debts for the three most populous states? Express your answer in scientific notation.

b. What is the sum of the debt for the three least populous states? Express your answer in scientific notation.

c. How much larger is the combined debt of the three most populous states than that of the three least populous states? Express your answer in scientific notation.

Exercise 8

a. What is the sum of the population of the three most populous states? Express your answer in scientific notation.

b. What is the sum of the population of the three least populous states? Express your answer in scientific notation.

c. Approximately how many times greater is the total population of California, New York, and Texas compared to the total population of North Dakota, Vermont, and Wyoming?

Exercise 9

All planets revolve around the sun in elliptical orbits. Uranus's furthest distance from the sun is approximately 3.004×10^9 km, and its closest distance is approximately 2.749×10^9 km. Using this information, what is the average distance of Uranus from the sun?

Problem Set

1. Write the number 68,127,000,000,000,000 in scientific notation. Which of the two representations of this number do you prefer? Explain.

2. Here are the masses of the so-called inner planets of the solar system.

   Mercury: 3.3022×10^{23} kg   Earth: 5.9722×10^{24} kg

   Venus: 4.8685×10^{24} kg   Mars: 6.4185×10^{23} kg

   What is the average mass of all four inner planets? Write your answer in scientific notation.

Lesson 10: Operations with Numbers in Scientific Notation

Exercise 1

The speed of light is 300,000,000 meters per second. The sun is approximately 1.5×10^{11} meters from Earth. How many seconds does it take for sunlight to reach Earth?

Exercise 2

The mass of the moon is about 7.3×10^{22} kg. It would take approximately 26,000,000 moons to equal the mass of the sun. Determine the mass of the sun.

Exercise 3

The mass of Earth is 5.9×10^{24} kg. The mass of Pluto is 13,000,000,000,000,000,000,000 kg. Compared to Pluto, how much greater is Earth's mass than Pluto's mass?

Exercise 4

Using the information in Exercises 2 and 3, find the combined mass of the moon, Earth, and Pluto.

Exercise 5

How many combined moon, Earth, and Pluto masses (i.e., the answer to Exercise 4) are needed to equal the mass of the sun (i.e., the answer to Exercise 2)?

Problem Set

1. The sun produces 3.8×10^{27} joules of energy per second. How much energy is produced in a year? (Note: a year is approximately 31,000,000 seconds).

2. On average, Mercury is about 57,000,000 km from the sun, whereas Neptune is about 4.5×10^9 km from the sun. What is the difference between Mercury's and Neptune's distances from the sun?

3. The mass of Earth is approximately 5.9×10^{24} kg, and the mass of Venus is approximately 4.9×10^{24} kg.

   a. Find their combined mass.

   b. Given that the mass of the sun is approximately 1.9×10^{30} kg, how many Venuses and Earths would it take to equal the mass of the sun?

1. The sun produces 3.8×10^{27} joules of energy per second. How much energy is produced in a year? (Note: a year is approximately 31,000,000 seconds).

2. On average, Mercury is about 57,000,000 km from the sun, whereas Neptune is about 4.5×10^9 km from the sun. What is the difference between Mercury's and Neptune's distances from the sun?

3. The mass of Earth is approximately 5.9×10^{24} kg, and the mass of Venus is approximately 4.9×10^{24} kg.

   a. Find their combined mass.

   b. Given that the mass of the sun is approximately 1.9×10^{30} kg, how many Venuses and Earths would it take to equal the mass of the sun?

Lesson 11: Efficacy of Scientific Notation

Exercise 1

The mass of a proton is

0.000 000 000 000 000 000 000 000 001 672 622 kg.

In scientific notation it is

Exercise 2

The mass of an electron is

0.000 000 000 000 000 000 000 000 000 910 938 291 kg.

In scientific notation it is

Exercise 3

Write the ratio that compares the mass of a proton to the mass of an electron.

Exercise 4

Compute how many times heavier a proton is than an electron (i.e., find the value of the ratio). Round your final answer to the nearest one.

Example 2

The U.S. national debt as of March 23, 2013, rounded to the nearest dollar, is $16,755,133,009,522. According to the 2012 U.S. census, there are about 313,914,040 U.S. citizens. What is each citizen's approximate share of the debt?

$$\frac{1.6755 \times 10^{13}}{3.14 \times 10^8} = \frac{1.6755}{3.14} \times \frac{10^{13}}{10^8}$$

$$= \frac{1.6755}{3.14} \times 10^5$$

$$= 0.533598\ldots \times 10^5$$

$$\approx 0.5336 \times 10^5$$

$$= 53360$$

Each U.S. citizen's share of the national debt is about $53,360.

Exercise 5

The geographic area of California is 163,696 sq. mi., and the geographic area of the U.S. is 3,794,101 sq. mi. Let's round off these figures to 1.637×10^5 and 3.794×10^6. In terms of area, roughly estimate how many Californias would make up one U.S. Then compute the answer to the nearest ones.

Exercise 6

The average distance from Earth to the moon is about 3.84×10^5 km, and the distance from Earth to Mars is approximately 9.24×10^7 km in year 2014. On this simplistic level, how much farther is traveling from Earth to Mars than from Earth to the moon?

Problem Set

1. There are approximately 7.5×10^{18} grains of sand on Earth. There are approximately 7×10^{27} atoms in an average human body. Are there more grains of sand on Earth or atoms in an average human body? How do you know?

2. About how many times more atoms are in a human body compared to grains of sand on Earth?

3. Suppose the geographic areas of California and the U.S. are 1.637×10^5 and 3.794×10^6 sq. mi., respectively. California's population (as of 2012) is approximately 3.804×10^7 people. If population were proportional to area, what would be the U.S. population?

4. The actual population of the U.S. (as of 2012) is approximately 3.14×10^8. How does the population density of California (i.e., the number of people per square mile) compare with the population density of the U.S.?

Lesson 12: Choice of Unit

Exercise 1

A certain brand of MP3 player will display how long it will take to play through its entire music library. If the maximum number of songs the MP3 player can hold is 1,000 (and the average song length is 4 minutes), would you want the time displayed in terms of seconds-, days-, or years-worth of music? Explain.

Exercise 2

You have been asked to make frosted cupcakes to sell at a school fundraiser. Each frosted cupcake contains about 20 grams of sugar. Bake sale coordinators expect 500 people will attend the event. Assume everyone who attends will buy a cupcake; does it make sense to buy sugar in grams, pounds, or tons? Explain.

Exercise 3

The seafloor spreads at a rate of approximately 10 cm per year. If you were to collect data on the spread of the seafloor each week, which unit should you use to record your data? Explain.

The gigaelectronvolt, $\dfrac{\text{GeV}}{c^2}$, is what particle physicists use as the unit of mass.

1 gigaelectronvolt = 1.783×10^{-27} kg

Mass of 1 proton = $1.672\,622 \times 10^{-27}$ kg

Exercise 4

Show that the mass of a proton is $0.938\dfrac{\text{GeV}}{c^2}$.

In popular science writing, a commonly used unit is the light-year, or the <u>distance</u> light travels in one year (note: one year is defined as 365.25 days).

1 light-year = $9,460,730,472,580.800$ km $\approx 9.46073 \times 10^{12}$ km

Exercise 5

The distance of the nearest star (Proxima Centauri) to the sun is approximately $4.013\,336\,473 \times 10^{13}$ km. Show that Proxima Centauri is 4.2421 light-years from the sun.

Exploratory Challenge 2

Suppose you are researching atomic diameters and find that credible sources provided the diameters of five different atoms as shown in the table below. All measurements are in centimeters.

1×10^{-8}	1×10^{-12}	5×10^{-8}	5×10^{-10}	5.29×10^{-11}

Exercise 6

What new unit might you introduce in order to discuss the differences in diameter measurements?

Exercise 7

Name your unit, and explain why you chose it.

Exercise 8

Using the unit you have defined, rewrite the five diameter measurements.

Problem Set

1. Verify the claim that, in terms of gigaelectronvolts, the mass of an electron is 0.000511.

2. The maximum distance between Earth and the sun is 1.52098232×10^8 km, and the minimum distance is 1.47098290×10^8 km.[1] What is the average distance between Earth and the sun in scientific notation?

3. Suppose you measure the following masses in terms of kilograms:

2.6×10^{21}	9.04×10^{23}
8.82×10^{23}	2.3×10^{18}
1.8×10^{12}	2.103×10^{22}
8.1×10^{20}	6.23×10^{18}
6.723×10^{19}	1.15×10^{20}
7.07×10^{21}	7.210×10^{29}
5.11×10^{25}	7.35×10^{24}
7.8×10^{19}	5.82×10^{26}

    What new unit might you introduce in order to aid discussion of the masses in this problem? Name your unit, and express it using some power of 10. Rewrite each number using your newly defined unit.

---

[1]Note: Earth's orbit is elliptical, not circular.

Lesson 13: Comparison of Numbers Written in Scientific Notation and Interpreting Scientific Notation Using Technology

Classwork

There is a general principle that underlies the comparison of two numbers in scientific notation: *Reduce everything to whole numbers if possible.* To this end, we recall two basic facts.

1. Inequality (A): Let x and y be numbers and let $z > 0$. Then $x < y$ if and only if $xz < yz$.
2. Comparison of whole numbers:
   a. If two whole numbers have different numbers of digits, then the one with more digits is greater.
   b. Suppose two whole numbers p and q have the same number of digits and, moreover, they agree digit-by-digit (starting from the left) until the nth place. If the digit of p in the $(n+1)^{\text{th}}$ place is greater than the corresponding digit in q, then $p > q$.

Exercise 1

The Fornax Dwarf galaxy is 4.6×10^5 light-years away from Earth, while Andromeda I is 2.430×10^6 light-years away from Earth. Which is closer to Earth?

Exercise 2

The average lifetime of the tau lepton is 2.906×10^{-13} seconds, and the average lifetime of the neutral pion is 8.4×10^{-17} seconds. Explain which subatomic particle has a longer average lifetime.

Exploratory Challenge 1/Exercise 3

THEOREM: *Given two positive numbers in scientific notation, $a \times 10^m$ and $b \times 10^n$, if $m < n$, then $a \times 10^m < b \times 10^n$.*

Prove the theorem.

Exercise 4

Compare 9.3×10^{28} and 9.2879×10^{28}.

Exercise 5

Chris said that $5.3 \times 10^{41} < 5.301 \times 10^{41}$ because 5.3 has fewer digits than 5.301. Show that even though his answer is correct, his reasoning is flawed. Show him an example to illustrate that his reasoning would result in an incorrect answer. Explain.

EUREKA MATH™ | Lesson 13:   Comparison of Numbers Written in Scientific Notation and Interpreting   S.64
                             Scientific Notation Using Technology

Exploratory Challenge 2/Exercise 6

You have been asked to determine the exact number of Google searches that are made each year. The only information you are provided is that there are 35,939,938,877 searches performed each week. Assuming the exact same number of searches are performed each week for the 52 weeks in a year, how many total searches will have been performed in one year? Your calculator does not display enough digits to get the exact answer. Therefore, you must break down the problem into smaller parts. Remember, you cannot approximate an answer because you need to find an exact answer. Use the screen shots below to help you reach your answer.

35 939 × 52 =

1868828

938877 × 52 =

48821604

Yahoo! is another popular search engine. Yahoo! receives requests for 1,792,671,355 searches each month. Assuming the same number of searches are performed each month, how many searches are performed on Yahoo! each year? Use the screen shots below to help determine the answer.

1792 x 12 =

21504

(	)	%	AC
7	8	9	÷
4	5	6	×
1	2	3	-
0	.	=	+

671 335 x 12 =

8056020

(	)	%	AC
7	8	9	÷
4	5	6	×
1	2	3	-
0	.	=	+

EUREKA MATH™

Lesson 13:   Comparison of Numbers Written in Scientific Notation and Interpreting
Scientific Notation Using Technology   S.66

Problem Set

1. Write out a detailed proof of the fact that, given two numbers in scientific notation, $a \times 10^n$ and $b \times 10^n$, $a < b$, if and only if $a \times 10^n < b \times 10^n$.

   a. Let A and B be two positive numbers, with no restrictions on their size. Is it true that $A \times 10^{-5} < B \times 10^5$?

   b. Now, if $A \times 10^{-5}$ and $B \times 10^5$ are written in scientific notation, is it true that $A \times 10^{-5} < B \times 10^5$? Explain.

2. The mass of a neutron is approximately 1.674927×10^{-27} kg. Recall that the mass of a proton is 1.672622×10^{-27} kg. Explain which is heavier.

3. The average lifetime of the Z boson is approximately 3×10^{-25} seconds, and the average lifetime of a neutral rho meson is approximately 4.5×10^{-24} seconds.

   a. Without using the theorem from today's lesson, explain why the neutral rho meson has a longer average lifetime.

   b. Approximately how much longer is the lifetime of a neutral rho meson than a Z boson?

Eureka Math
Grade 8
Module 2

Special thanks go to the Gordan A. Cain Center and to the Department of Mathematics at Louisiana State University for their support in the development of *Eureka Math*.

Published by Great Minds

Printed in the U.S.A.

This book may be purchased from the publisher at eureka-math.org

10 9 8 7 6 5 4 3 2 1

ISBN 978-1-63255-320-1

Lesson 1: Why Move Things Around?

Exploratory Challenge

a.   Describe, intuitively, what kind of transformation is required to move the figure on the left to each of the figures (1)–(3) on the right. To help with this exercise, use a transparency to copy the figure on the left. Note: Begin by moving the left figure to each of the locations in (1), (2), and (3).

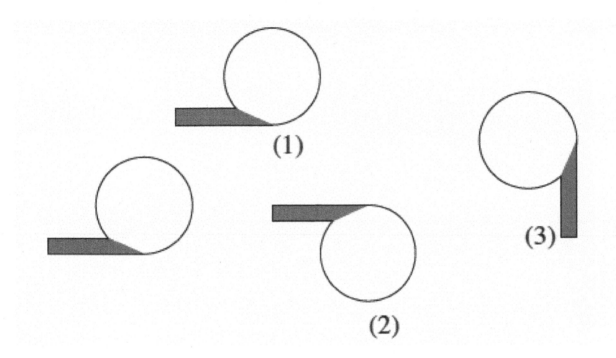

b. Given two segments AB and CD, which could be very far apart, how can we find out if they have the same length without measuring them individually? Do you think they have the same length? How do you check? In other words, why do you think we need to move things around on the plane?

Lesson Summary

A *transformation* F of the plane is a function that assigns to each point P of the plane a point $F(P)$ in the plane.

- By definition, the symbol $F(P)$ denotes a specific single point, unambiguously.
- The point $F(P)$ will be called the image of P by F. Sometimes the image of P by F is denoted simply as P' (read "P prime").
- The transformation F is sometimes said to "move" the point P to the point $F(P)$.
- We also say F maps P to $F(P)$.

In this module, we will mostly be interested in transformations that are given by rules, that is, a set of step-by-step instructions that can be applied to any point P in the plane to get its image.

If given any two points P and Q, the distance between the images $F(P)$ and $F(Q)$ is the same as the distance between the original points P and Q, and then the transformation F preserves distance, or is distance-preserving.

- A distance-preserving transformation is called a *rigid motion* (or an *isometry*), and the name suggests that it moves the points of the plane around in a rigid fashion.

Problem Set

1. Using as much of the new vocabulary as you can, try to describe what you see in the diagram below.

2. Describe, intuitively, what kind of transformation is required to move Figure A on the left to its image on the right.

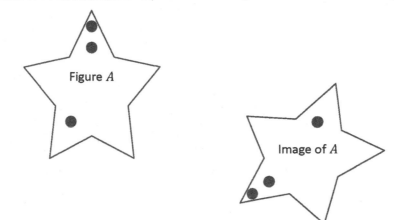

A transformation _F_ of the plane is a function that assigns to each point _P_ of the plane a point _F(P)_ in the plane.

- By definition, the symbol _F(P)_ denotes a specific single point, unambiguously.
- The point _F(P)_ will be called the image of _P_ by _F_. Sometimes the image of _P_ is denoted simply as _P'_ (read "_P_ prime").
- The transformation _F_ is sometimes said to "move" the point _P_ to the point _F(P)_.
- We also say _F_ maps _P_ to _F(P)_.

In this lesson, we will be interested in transformations that are given by rules, that is, a set of step-by-step instructions that can be applied to any point _P_ in the plane to get its image.

If given any two points _P_ and _Q_, the distance between the images _F(P)_ and _F(Q)_ is the same as the distance between _P_ and _Q_, then the transformation preserves distance, or is distance-preserving.

- A distance-preserving transformation is called a rigid motion (or isometry), and the name suggests that it moves the points of the plane around in a rigid fashion.

1. Using as much of the new vocabulary as you can, try to describe what you see in the diagram below.

_F(A)_

_F(S)_

2. Describe, intuitively, what kind of transformation is required to move Figure A on the left to its image on the right.

Figure A

Image of A

Lesson 2: Definition of Translation and Three Basic Properties

Exercise 1

Draw at least three different vectors, and show what a translation of the plane along each vector looks like. Describe what happens to the following figures under each translation using appropriate vocabulary and notation as needed.

Exercise 2

The diagram below shows figures and their images under a translation along $\overrightarrow{HI}$. Use the original figures and the translated images to fill in missing labels for points and measures.

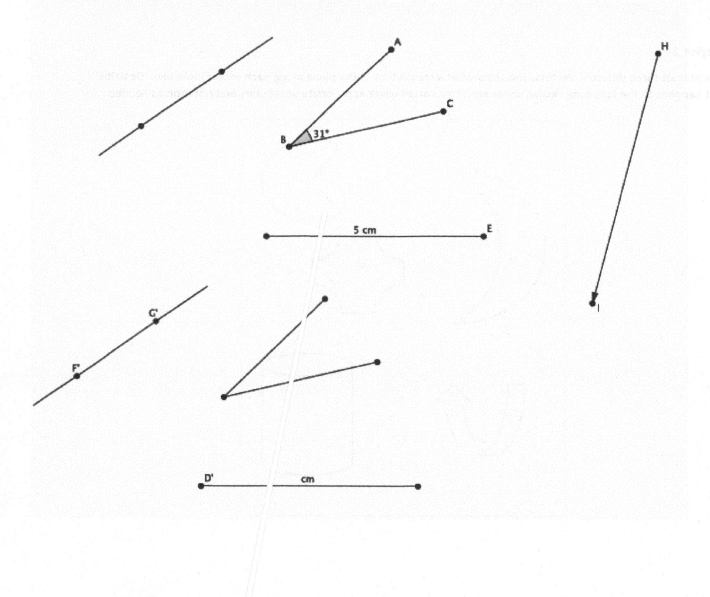

Lesson Summary

Translation occurs along a given vector:

- A *vector* is directed line segment, that is, it is a segment with a direction given by connecting one of its endpoint (called the *initial point* or *starting point*) to the other endpoint (called the *terminal point* or simply the *endpoint*). It is often represented as an "arrow" with a "tail" and a "tip."

- The *length of a vector* is, by definition, the length of its underlying segment.

- Pictorially note the starting and endpoints:

A translation of a plane along a given vector is a basic rigid motion of a plane.

The three basic properties of translation are as follows:

(Translation 1) A translation maps a line to a line, a ray to a ray, a segment to a segment, and an angle to an angle.

(Translation 2) A translation preserves lengths of segments.

(Translation 3) A translation preserves measures of angles.

Terminology

TRANSLATION (description): For vector $\overrightarrow{AB}$, a *translation along* $\overrightarrow{AB}$ is the transformation of the plane that maps each point C of the plane to its image C' so that the line $\overleftrightarrow{CC'}$ is parallel to the vector (or contains it), and the vector $\overrightarrow{CC'}$ points in the same direction and is the same length as the vector $\overrightarrow{AB}$.

Lesson 2:   Definition of Translation and Three Basic Properties   S.7

Problem Set

1. Translate the plane containing Figure A along $\overrightarrow{AB}$. Use your transparency to sketch the image of Figure A by this translation. Mark points on Figure A, and label the image of Figure A accordingly.

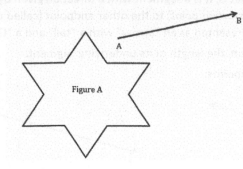

2. Translate the plane containing Figure B along $\overrightarrow{BA}$. Use your transparency to sketch the image of Figure B by this translation. Mark points on Figure B, and label the image of Figure B accordingly.

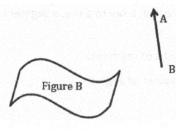

3. Draw an acute angle (your choice of degree), a segment with length 3 cm, a point, a circle with radius 1 in., and a vector (your choice of length, i.e., starting point and ending point). Label points and measures (measurements do not need to be precise, but your figure must be labeled correctly). Use your transparency to translate all of the figures you have drawn along the vector. Sketch the images of the translated figures and label them.

4. What is the length of the translated segment? How does this length compare to the length of the original segment? Explain.

5. What is the length of the radius in the translated circle? How does this radius length compare to the radius of the original circle? Explain.

6. What is the degree of the translated angle? How does this degree compare to the degree of the original angle? Explain.

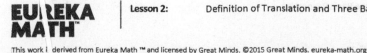

7. Translate point D along vector $\overrightarrow{AB}$, and label the image D'. What do you notice about the line containing vector $\overrightarrow{AB}$ and the line containing points D and D'? (Hint: Will the lines ever intersect?)

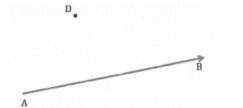

8. Translate point E along vector $\overrightarrow{AB}$, and label the image E'. What do you notice about the line containing vector $\overrightarrow{AB}$ and the line containing points E and E'?

Lesson 3: Translating Lines

Classwork

Exercises

1. Draw a line passing through point P that is parallel to line L. Draw a second line passing through point P that is parallel to line L and that is distinct (i.e., different) from the first one. What do you notice?

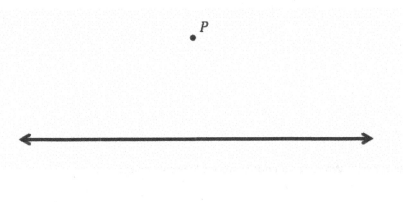

2. Translate line L along the vector $\vec{AB}$. What do you notice about L and its image, L'?

3.   Line L is parallel to vector $\overrightarrow{AB}$. Translate line L along vector $\overrightarrow{AB}$. What do you notice about L and its image, L'?

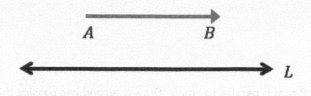

4.   Translate line L along the vector $\overrightarrow{AB}$. What do you notice about L and its image, L'?

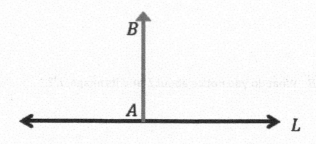

5. Line L has been translated along vector $\overrightarrow{AB}$, resulting in L'. What do you know about lines L and L'?

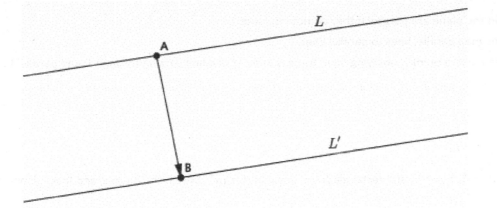

6. Translate L_1 and L_2 along vector $\overrightarrow{DE}$. Label the images of the lines. If lines L_1 and L_2 are parallel, what do you know about their translated images?

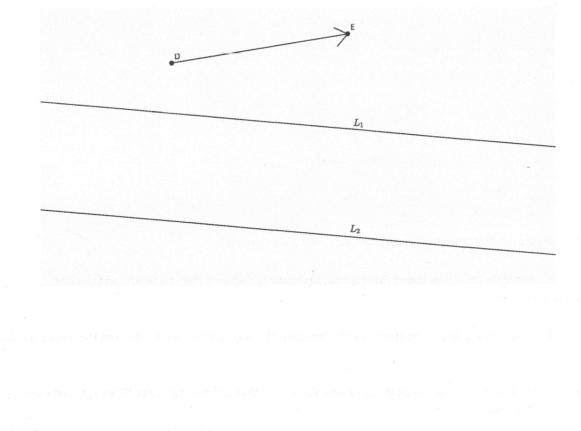

Lesson Summary

- Two lines in the plane are parallel if they do not intersect.
- Translations map parallel lines to parallel lines.
- Given a line L and a point P not lying on L, there is at most one line passing through P and parallel to L.

Problem Set

1. Translate $\angle XYZ$, point A, point B, and rectangle $HIJK$ along vector $\overrightarrow{EF}$. Sketch the images, and label all points using prime notation.

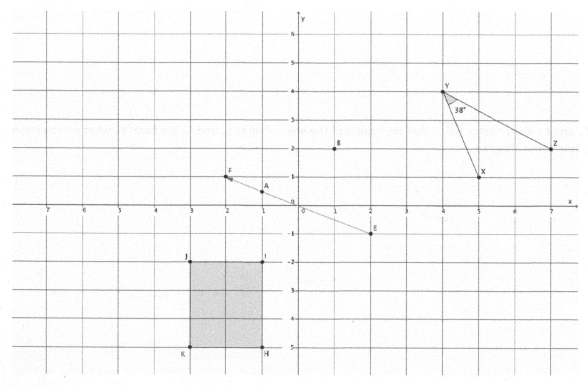

2. What is the measure of the translated image of $\angle XYZ$? How do you know?

3. Connect B to B'. What do you know about the line that contains the segment formed by BB' and the line containing the vector $\overrightarrow{EF}$?

4. Connect A to A'. What do you know about the line that contains the segment formed by AA' and the line containing the vector $\overrightarrow{EF}$?

5. Given that figure $HIJK$ is a rectangle, what do you know about lines that contain segments HI and JK and their translated images? Explain.

Lesson 3:	Translating Lines

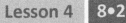

Lesson 4: Definition of Reflection and Basic Properties

Classwork

Exercises

1. Reflect △ ABC and Figure D across line L. Label the reflected images.

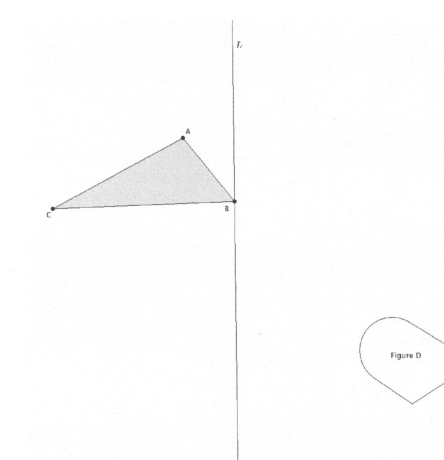

2. Which figure(s) were not moved to a new location on the plane under this transformation?

3. Reflect the images across line L. Label the reflected images.

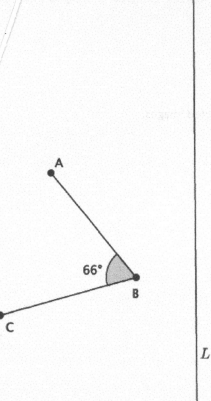

4. Answer the questions about the image above.

    a. Use a protractor to measure the reflected $\angle ABC$. What do you notice?

    b. Use a ruler to measure the length of IJ and the length of the image of IJ after the reflection. What do you notice?

5.   Reflect Figure R and $\triangle EFG$ across line L. Label the reflected images.

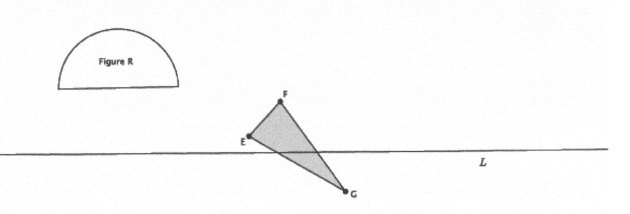

Basic Properties of Reflections:

   (Reflection 1) A reflection maps a line to a line, a ray to a ray, a segment to a segment, and an angle to an angle.

   (Reflection 2) A reflection preserves lengths of segments.

   (Reflection 3) A reflection preserves measures of angles.

If the reflection is across a line L and P is a point not on L, then L bisects and is perpendicular to the segment PP', joining P to its reflected image P'. That is, the lengths of OP and OP' are equal.

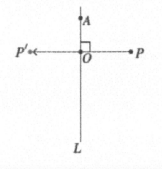

Use the picture below for Exercises 6–9.

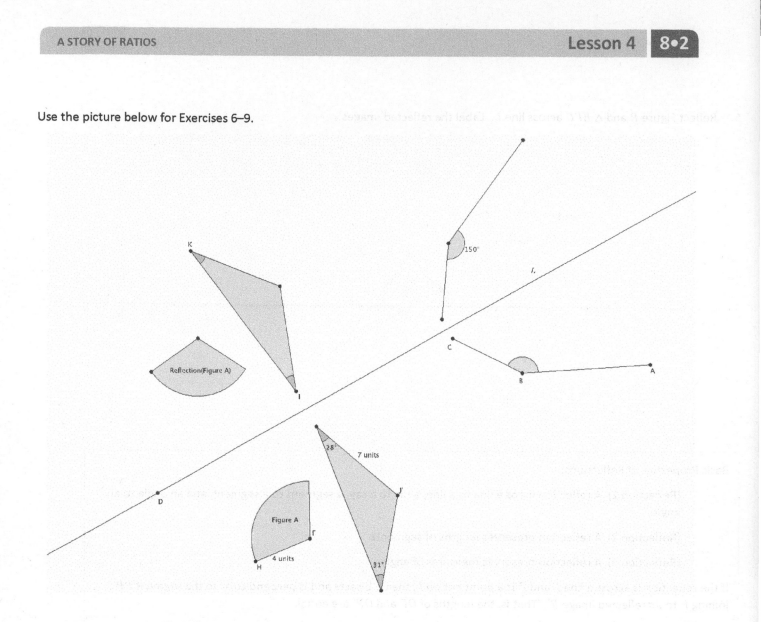

6. Use the picture to label the unnamed points.

7. What is the measure of ∠JKI? ∠KIJ? ∠ABC? How do you know?

8. What is the length of segment Reflection(FH)? IJ? How do you know?

9. What is the location of Reflection(D)? Explain.

Lesson Summary

- A reflection is another type of basic rigid motion.

- A reflection across a line maps one half-plane to the other half-plane; that is, it maps points from one side of the line to the other side of the line. The reflection maps each point on the line to itself. The line being reflected across is called the *line of reflection*.

- When a point P is joined with its reflection P' to form the segment PP', the line of reflection bisects and is perpendicular to the segment PP'.

Terminology

REFLECTION (description): Given a line L in the plane, a *reflection across L* is the transformation of the plane that maps each point on the line L to itself, and maps each remaining point P of the plane to its image P' such that L is the perpendicular bisector of the segment PP'.

Problem Set

1. In the picture below, $\angle DEF = 56°$, $\angle ACB = 114°$, $AB = 12.6$ units, $JK = 5.32$ units, point E is on line L, and point I is off of line L. Let there be a reflection across line L. Reflect and label each of the figures, and answer the questions that follow.

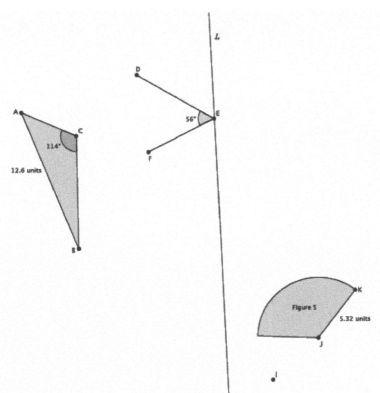

2. What is the measure of $Reflection(\angle DEF)$? Explain.

3. What is the length of $Reflection(JK)$? Explain.

4. What is the measure of $Reflection(\angle ACB)$?

5. What is the length of $Reflection(AB)$?

6. Two figures in the picture were not moved under the reflection. Name the two figures, and explain why they were not moved.

7. Connect points I and I'. Name the point of intersection of the segment with the line of reflection point Q. What do you know about the lengths of segments IQ and QI'?

Lesson 5: Definition of Rotation and Basic Properties

Classwork

Exercises

1. Let there be a rotation of d degrees around center O. Let P be a point other than O. Select d so that $d \geq 0$. Find P' (i.e., the rotation of point P) using a transparency.

2. Let there be a rotation of d degrees around center O. Let P be a point other than O. Select d so that $d < 0$. Find P' (i.e., the rotation of point P) using a transparency.

3. Which direction did the point P rotate when $d \geq 0$?

4. Which direction did the point P rotate when $d < 0$?

5. Let L be a line, $\overrightarrow{AB}$ be a ray, $\overline{CD}$ be a segment, and $\angle EFG$ be an angle, as shown. Let there be a rotation of d degrees around point O. Find the images of all figures when $d \geq 0$.

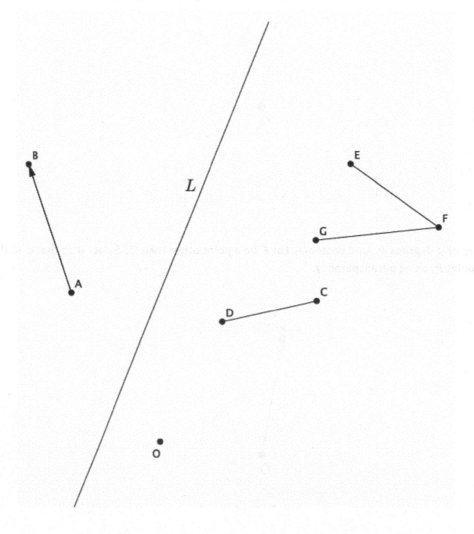

6. Let $\overline{AB}$ be a segment of length 4 units and $\angle CDE$ be an angle of size 45°. Let there be a rotation by d degrees, where $d < 0$, about O. Find the images of the given figures. Answer the questions that follow.

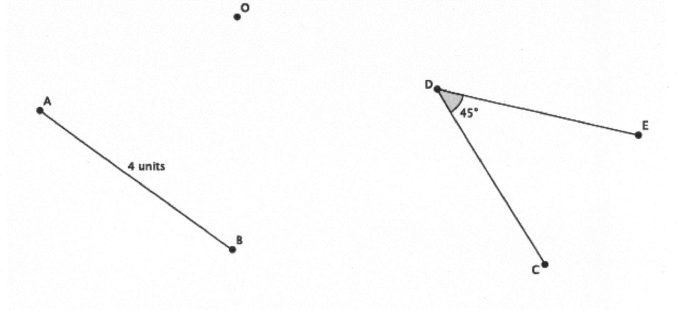

a. What is the length of the rotated segment $Rotation(AB)$?

b. What is the degree of the rotated angle $Rotation(\angle CDE)$?

7. Let L_1 and L_2 be parallel lines. Let there be a rotation by d degrees, where $-360 < d < 360$, about O. Is $(L_1)' \parallel (L_2)'$?

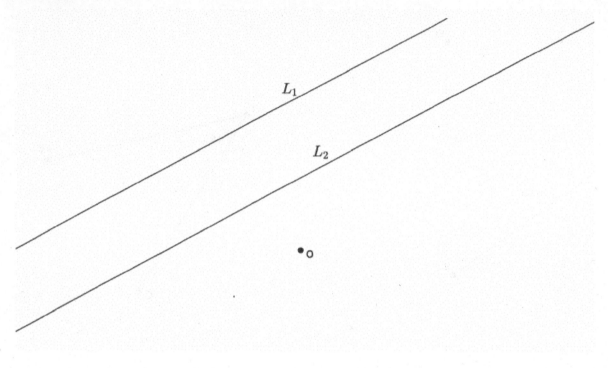

8. Let L be a line and O be the center of rotation. Let there be a rotation by d degrees, where $d \neq 180$ about O. Are the lines L and L' parallel?

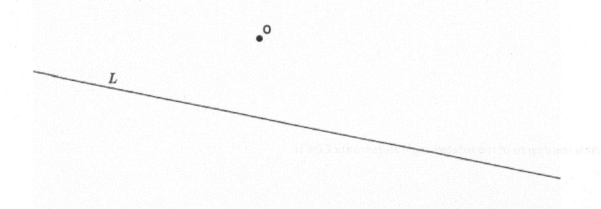

Lesson Summary

Rotations require information about the center of rotation and the degree in which to rotate. Positive degrees of rotation move the figure in a counterclockwise direction. Negative degrees of rotation move the figure in a clockwise direction.

Basic Properties of Rotations:

- (Rotation 1) A rotation maps a line to a line, a ray to a ray, a segment to a segment, and an angle to an angle.
- (Rotation 2) A rotation preserves lengths of segments.
- (Rotation 3) A rotation preserves measures of angles.

When parallel lines are rotated, their images are also parallel. A line is only parallel to itself when rotated exactly 180°.

Terminology

ROTATION (DESCRIPTION): For a number d between 0 and 180, the *rotation of d degrees around center O* is the transformation of the plane that maps the point O to itself, and maps each remaining point P of the plane to its image P' in the counterclockwise half-plane of ray $\overrightarrow{OP}$ so that P and P' are the same distance away from O and the measurement of $\angle P'OP$ is d degrees.

The *counterclockwise half-plane* is the half-plane that lies to the left of $\overrightarrow{OP}$ while moving along $\overrightarrow{OP}$ in the direction from O to P.

Problem Set

1. Let there be a rotation by $-90°$ around the center O.

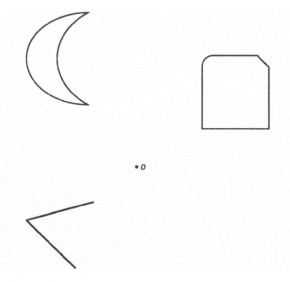

2. Explain why a rotation of 90 degrees around any point O never maps a line to a line parallel to itself.

3. A segment of length 94 cm has been rotated d degrees around a center O. What is the length of the rotated segment? How do you know?

4. An angle of size 124° has been rotated d degrees around a center O. What is the size of the rotated angle? How do you know?

Lesson 6: Rotations of 180 Degrees

Classwork

Example 1

The picture below shows what happens when there is a rotation of 180° around center O.

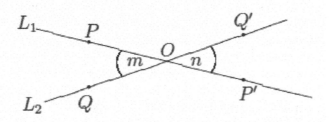

Example 2

The picture below shows what happens when there is a rotation of 180° around center O, the origin of the coordinate plane.

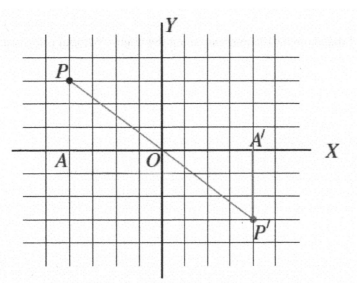

Exercises 1–9

1. Using your transparency, rotate the plane 180 degrees, about the origin. Let this rotation be $Rotation_0$. What are the coordinates of $Rotation_0(2, -4)$?

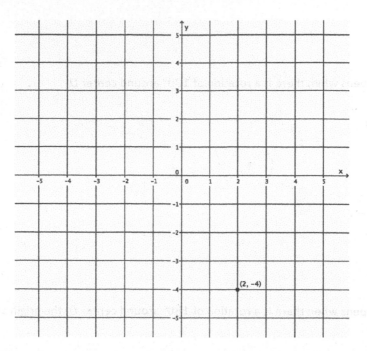

2. Let $Rotation_0$ be the rotation of the plane by 180 degrees, about the origin. <u>Without</u> using your transparency, find $Rotation_0(-3, 5)$.

3. Let $Rotation_0$ be the rotation of 180 degrees around the origin. Let L be the line passing through $(-6, 6)$ parallel to the x-axis. Find $Rotation_0(L)$. Use your transparency if needed.

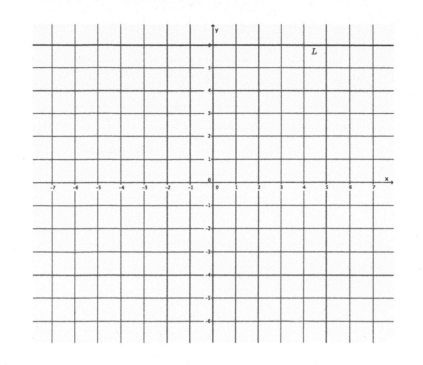

4. Let $Rotation_0$ be the rotation of 180 degrees around the origin. Let L be the line passing through $(7,0)$ parallel to the y-axis. Find $Rotation_0(L)$. Use your transparency if needed.

5. Let $Rotation_0$ be the rotation of 180 degrees around the origin. Let L be the line passing through $(0,2)$ parallel to the x-axis. Is L parallel to $Rotation_0(L)$?

6. Let $Rotation_0$ be the rotation of 180 degrees around the origin. Let L be the line passing through $(4,0)$ parallel to the y-axis. Is L parallel to $Rotation_0(L)$?

7. Let $Rotation_0$ be the rotation of 180 degrees around the origin. Let L be the line passing through $(0, -1)$ parallel to the x-axis. Is L parallel to $Rotation_0(L)$?

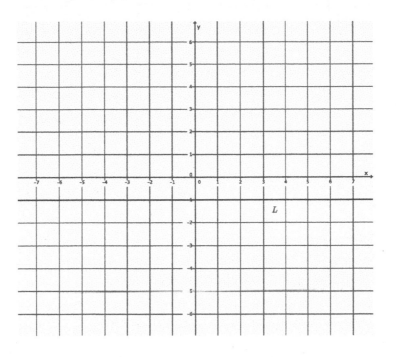

8. Let $Rotation_0$ be the rotation of 180 degrees around the origin. Is L parallel to $Rotation_0(L)$? Use your transparency if needed.

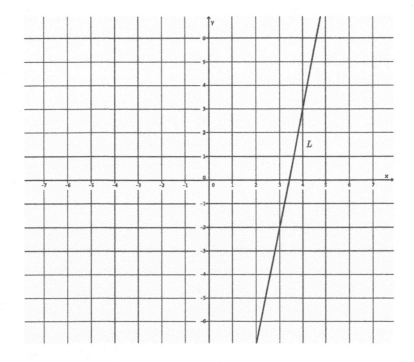

EUREKA MATH™

9. Let $Rotation_0$ be the rotation of 180 degrees around the center O. Is L parallel to $Rotation_0(L)$? Use your transparency if needed.

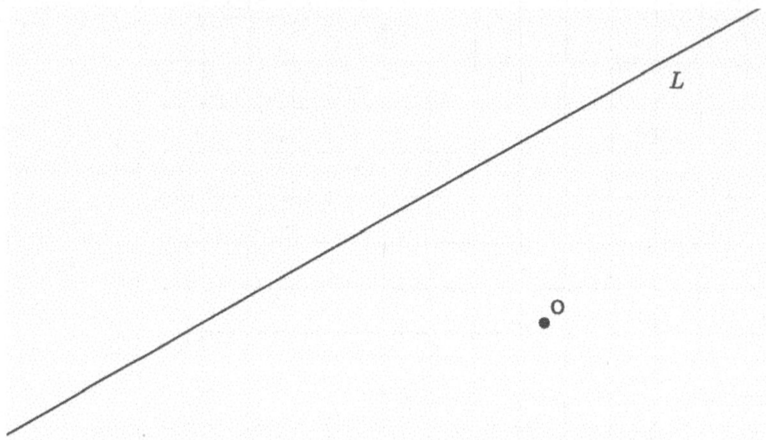

L

O

Lesson Summary

- A rotation of 180 degrees around O is the rigid motion so that if P is any point in the plane, P, O, and $Rotation(P)$ are *collinear* (i.e., lie on the same line).

- Given a 180-degree rotation, R_0 around the origin O of a coordinate system, R_0, and a point P with coordinates (a, b), it is generally said that $R_0(P)$ is the point with coordinates $(-a, -b)$.

THEOREM: Let O be a point not lying on a given line L. Then, the 180-degree rotation around O maps L to a line parallel to L.

Problem Set

Use the following diagram for Problems 1–5. Use your transparency as needed.

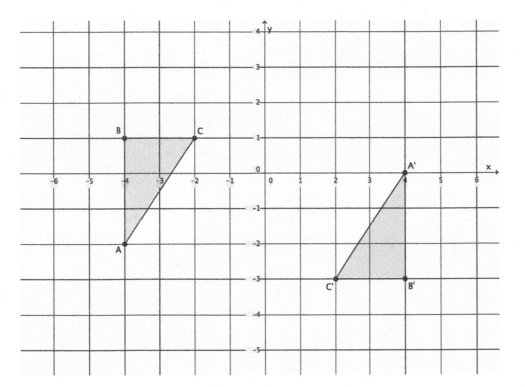

1. Looking only at segment BC, is it possible that a 180° rotation would map segment BC onto segment $B'C'$? Why or why not?

2. Looking only at segment AB, is it possible that a 180° rotation would map segment AB onto segment $A'B'$? Why or why not?

3. Looking only at segment AC, is it possible that a 180° rotation would map segment AC onto segment $A'C'$? Why or why not?

4. Connect point B to point B', point C to point C', and point A to point A'. What do you notice? What do you think that point is?

5. Would a rotation map triangle ABC onto triangle $A'B'C'$? If so, define the rotation (i.e., degree and center). If not, explain why not.

6. The picture below shows right triangles ABC and $A'B'C'$, where the right angles are at B and B'. Given that $AB = A'B' = 1$, and $BC = B'C' = 2$, and that $\overline{AB}$ is not parallel to $\overline{A'B'}$, is there a 180° rotation that would map △ ABC onto △ $A'B'C'$? Explain.

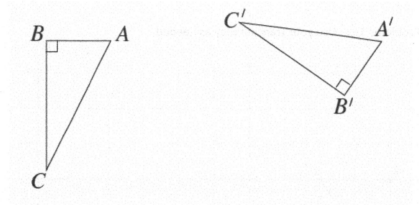

Lesson 7: Sequencing Translations

Classwork

Exploratory Challenge/Exercises 1–4

1.

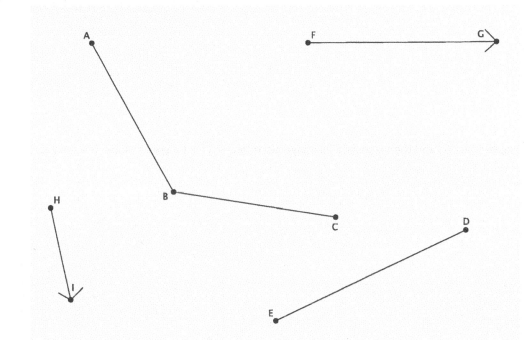

a.   Translate $\angle ABC$ and segment ED along vector $\overrightarrow{FG}$. Label the translated images appropriately, that is, $\angle A'B'C'$ and segment $E'D'$.

b.   Translate $\angle A'B'C'$ and segment $E'D'$ along vector $\overrightarrow{HI}$. Label the translated images appropriately, that is, $\angle A''B''C''$ and segment $E''D''$.

c.   How does the size of $\angle ABC$ compare to the size of $\angle A''B''C''$?

d.   How does the length of segment ED compare to the length of the segment $E''D''$?

e.   Why do you think what you observed in parts (d) and (e) were true?

2.   Translate $\triangle ABC$ along vector $\overrightarrow{FG}$, and then translate its image along vector $\overrightarrow{JK}$. Be sure to label the images appropriately.

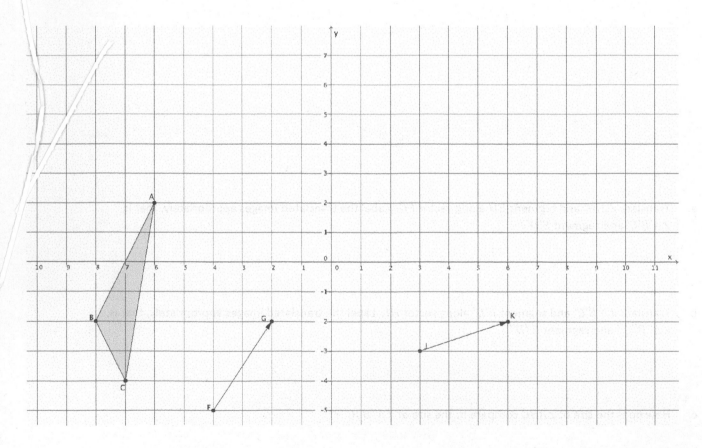

3. Translate figure $ABCDEF$ along vector $\overrightarrow{GH}$. Then translate its image along vector $\overrightarrow{JI}$. Label each image appropriately.

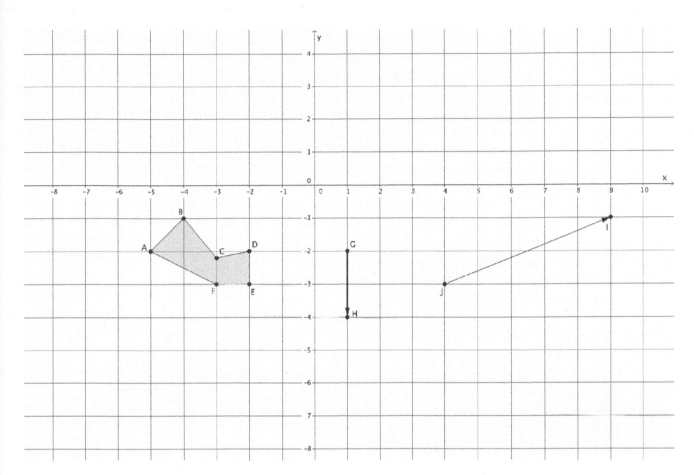

4.

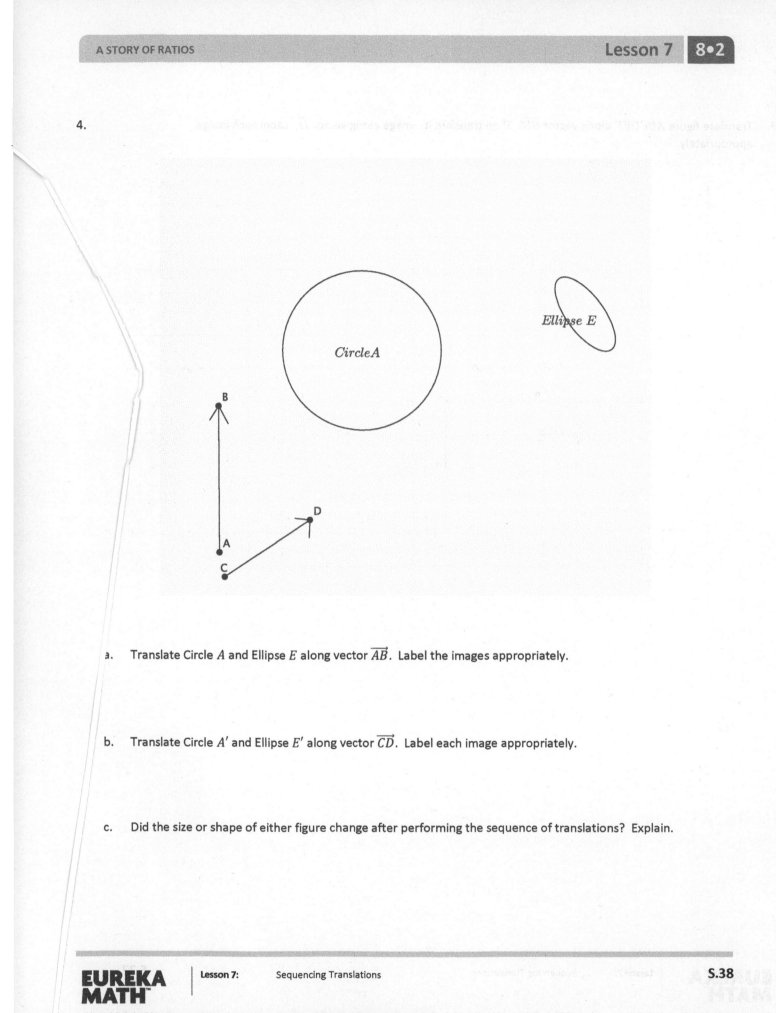

Circle A

Ellipse E

a. Translate Circle A and Ellipse E along vector $\overrightarrow{AB}$. Label the images appropriately.

b. Translate Circle A' and Ellipse E' along vector $\overrightarrow{CD}$. Label each image appropriately.

c. Did the size or shape of either figure change after performing the sequence of translations? Explain.

5. The picture below shows the translation of Circle A along vector $\overrightarrow{CD}$. Name the vector that maps the image of Circle A back to its original position.

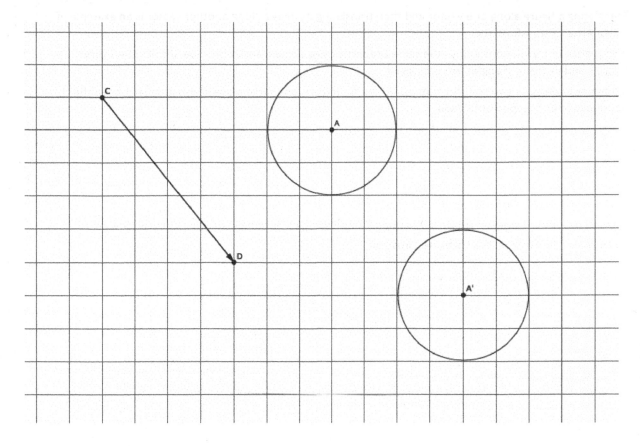

6. If a figure is translated along vector $\overrightarrow{QR}$, what translation takes the figure back to its original location?

Lesson Summary

- Translating a figure along one vector and then translating its image along another vector is an example of a sequence of transformations.

- A sequence of translations enjoys the same properties as a single translation. Specifically, the figures' lengths and degrees of angles are preserved.

- If a figure undergoes two transformations, F and G, and is in the same place it was originally, then the figure has been mapped onto itself.

Problem Set

1. Sequence translations of parallelogram $ABCD$ (a quadrilateral in which both pairs of opposite sides are parallel) along vectors $\overrightarrow{HG}$ and $\overrightarrow{FE}$. Label the translated images.

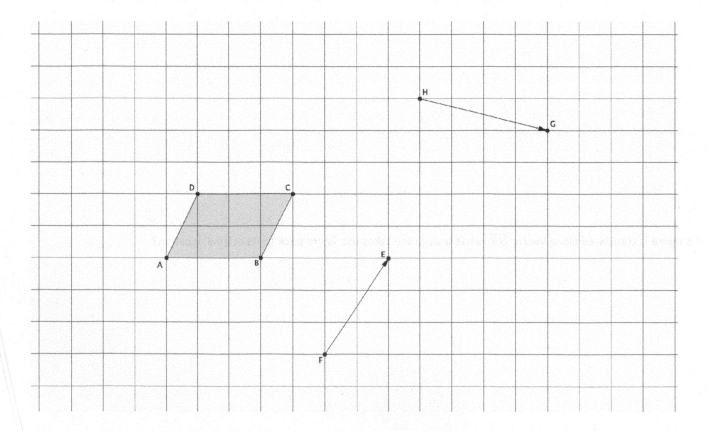

2. What do you know about $\overline{AD}$ and $\overline{BC}$ compared with $\overline{A'D'}$ and $\overline{B'C'}$? Explain.

3. Are the segments $A'B'$ and $A''B''$ equal in length? How do you know?

4. Translate the curved shape ABC along the given vector. Label the image.

5. What vector would map the shape $A'B'C'$ back onto shape ABC?

Lesson 8: Sequencing Reflections and Translations

Exercises 1–3

Use the figure below to answer Exercises 1–3.

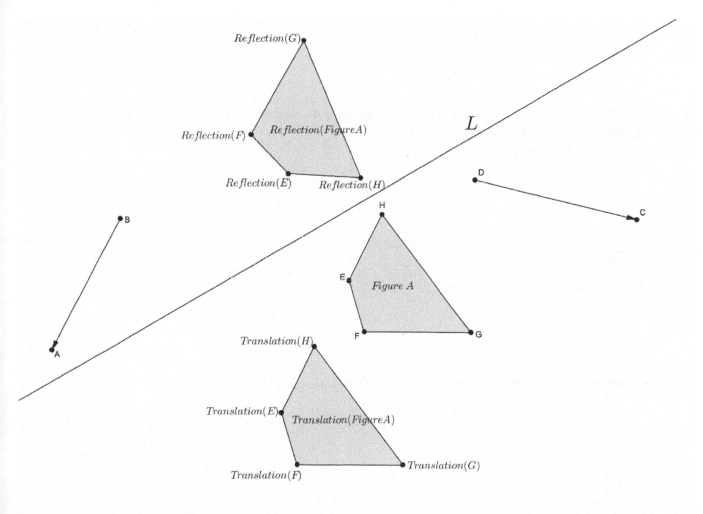

1. Figure A was translated along vector $\overrightarrow{BA}$, resulting in $Translation(Figure\ A)$. Describe a sequence of translations that would map Figure A back onto its original position.

2. Figure A was reflected across line L, resulting in $Reflection(Figure\ A)$. Describe a sequence of reflections that would map Figure A back onto its original position.

3. Can $Translation_{\overline{BA}}$ of Figure A undo the transformation of $Translation_{\overline{DC}}$ of Figure A? Why or why not?

Exercises 4–7

Let S be the black figure.

L

A

B

4. Let there be the translation along vector $\overrightarrow{AB}$ and a reflection across line L.

Use a transparency to perform the following sequence: Translate figure S; then, reflect figure S. Label the image S'.

5. Let there be the translation along vector $\overrightarrow{AB}$ and a reflection across line L.

Use a transparency to perform the following sequence: Reflect figure S; then, translate figure S. Label the image S''.

6. Using your transparency, show that under a sequence of any two translations, $Translation$ and $Translation_0$ (along different vectors), that the sequence of the $Translation$ followed by the $Translation_0$ is equal to the sequence of the $Translation_0$ followed by the $Translation$. That is, draw a figure, A, and two vectors. Show that the translation along the first vector, followed by a translation along the second vector, places the figure in the same location as when you perform the translations in the reverse order. (This fact is proven in high school Geometry.) Label the transformed image A'. Now, draw two new vectors and translate along them just as before. This time, label the transformed image A''. Compare your work with a partner. Was the statement "the sequence of the $Translation$ followed by the $Translation_0$ is equal to the sequence of the $Translation_0$ followed by the $Translation$" true in all cases? Do you think it will always be true?

7. Does the same relationship you noticed in Exercise 6 hold true when you replace one of the translations with a reflection. That is, is the following statement true: A translation followed by a reflection is equal to a reflection followed by a translation?

Lesson Summary

- A reflection across a line followed by a reflection across the same line places all figures in the plane back onto their original position.

- A reflection followed by a translation does not necessarily place a figure in the same location in the plane as a translation followed by a reflection. The order in which we perform a sequence of rigid motions matters.

Problem Set

1. Let there be a reflection across line L, and let there be a translation along vector $\overrightarrow{AB}$, as shown. If S denotes the black figure, compare the translation of S followed by the reflection of S with the reflection of S followed by the translation of S.

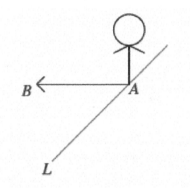

2. Let L_1 and L_2 be parallel lines, and let $Reflection_1$ and $Reflection_2$ be the reflections across L_1 and L_2, respectively (in that order). Show that a $Reflection_2$ followed by $Reflection_1$ is not equal to a $Reflection_1$ followed by $Reflection_2$. (Hint: Take a point on L_1 and see what each of the sequences does to it.)

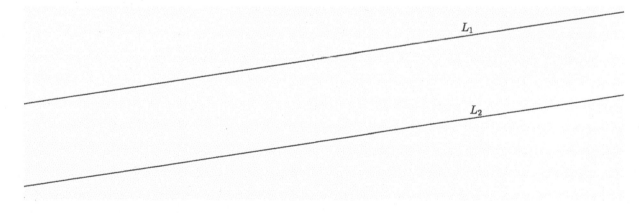

3. Let L_1 and L_2 be parallel lines, and let $Reflection_1$ and $Reflection_2$ be the reflections across L_1 and L_2, respectively (in that order). Can you guess what $Reflection_1$ followed by $Reflection_2$ is? Give as persuasive an argument as you can. (Hint: Examine the work you just finished for the last problem.)

- A reflection across a line followed by a reflection across the same line places all figures in the plane back to their original position.

- A reflection followed by a translation does not necessarily place a figure in the same location in the plane as a translation followed by a reflection. The order in which we perform a sequence of rigid motions matters.

1. Let there be a reflection across line ℓ, let there be a translation along vector $\vec{HI}$, as shown. If S denotes the black figure, compute the transformation of the figure by the reflection and then the reflection of S followed by the translation of.

2. Let L_1 and L_2 be parallel lines, and let Ref(lection)$_1$ and Ref(lection)$_2$ be the reflections across L_1 and L_2 respectively (in that order). Show that a Ref(lection)$_1$ followed by Ref(lection)$_2$ is not equal to a Ref(lection)$_2$ followed by Ref(lection)$_1$. [Hint: Take a point on L_2 and see what each of the sequences does to it.]

3. Let L_1 and L_2 be parallel lines, and let Ref(lection)$_1$ and Ref(lection)$_2$ be the reflections across L_1 and L_2 respectively (in that order). Can you guess what Ref(lection)$_1$ followed by Ref(lection)$_2$ is? Give as persuasive an argument as you can. [Hint: Examine the work you just finished for the last problem.]

Lesson 9: Sequencing Rotations

Classwork

Exploratory Challenge

1.

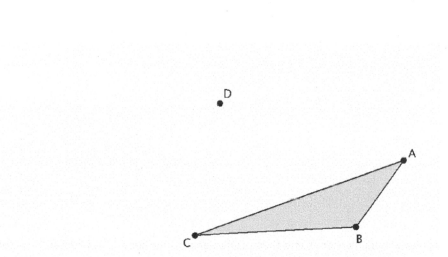

a.   Rotate △ ABC d degrees around center D. Label the rotated image as △ $A'B'C'$.

b.   Rotate △ $A'B'C'$ d degrees around center E. Label the rotated image as △ $A''B''C''$.

c.   Measure and label the angles and side lengths of △ ABC. How do they compare with the images △ $A'B'C'$ and △ $A''B''C''$?

d.   How can you explain what you observed in part (c)? What statement can you make about properties of sequences of rotations as they relate to a single rotation?

2.

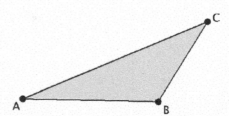

a. Rotate △ ABC d degrees around center D, and then rotate again d degrees around center E. Label the image as △ $A'B'C'$ after you have completed both rotations.

b. Can a single rotation around center D map △ $A'B'C'$ onto △ ABC?

c. Can a single rotation around center E map △ $A'B'C'$ onto △ ABC?

d. Can you find a center that would map △ $A'B'C'$ onto △ ABC in one rotation? If so, label the center F.

3.

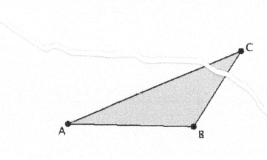

a.   Rotate △ ABC 90° (counterclockwise) around center D, and then rotate the image another 90° (counterclockwise) around center E. Label the image △ $A'B'C'$.

b.   Rotate △ ABC 90° (counterclockwise) around center E, and then rotate the image another 90° (counterclockwise) around center D. Label the image △ $A''B''C''$.

c.   What do you notice about the locations of △ $A'B'C'$ and △ $A''B''C''$? Does the order in which you rotate a figure around different centers have an impact on the final location of the figure's image?

4.

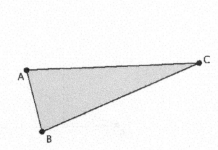

a. Rotate △ ABC 90° (counterclockwise) around center D, and then rotate the image another 45° (counterclockwise) around center D. Label the image △ $A'B'C'$.

b. Rotate △ ABC 45° (counterclockwise) around center D, and then rotate the image another 90° (counterclockwise) around center D. Label the image △ $A''B''C''$.

c. What do you notice about the locations of △ $A'B'C'$ and △ $A''B''C''$? Does the order in which you rotate a figure around the same center have an impact on the final location of the figure's image?

5. $\triangle ABC$ has been rotated around two different centers, and its image is $\triangle A'B'C'$. Describe a sequence of rigid motions that would map $\triangle ABC$ onto $\triangle A'B'C'$.

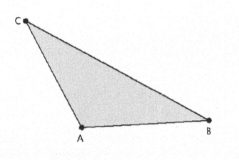
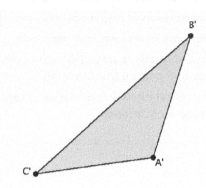

Lesson Summary

- Sequences of rotations have the same properties as a single rotation:
  - A sequence of rotations preserves degrees of measures of angles.
  - A sequence of rotations preserves lengths of segments.
- The order in which a sequence of rotations around different centers is performed matters with respect to the final location of the image of the figure that is rotated.
- The order in which a sequence of rotations around the same center is performed does not matter. The image of the figure will be in the same location.

Problem Set

1. Refer to the figure below.

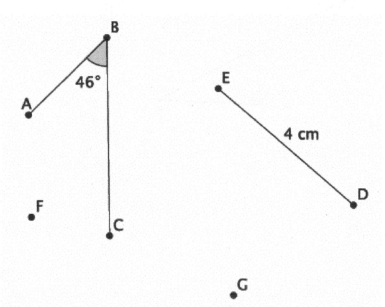

a. Rotate ∠ABC and segment DE d degrees around center F and then d degrees around center G. Label the final location of the images as ∠A'B'C' and segment D'E'.

b. What is the size of ∠ABC, and how does it compare to the size of ∠A'B'C'? Explain.

c. What is the length of segment DE, and how does it compare to the length of segment D'E'? Explain.

2. Refer to the figure given below.

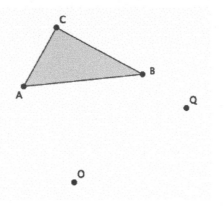

a. Let $Rotation_1$ be a counterclockwise rotation of 90° around the center O. Let $Rotation_2$ be a clockwise rotation of $(-45)°$ around the center Q. Determine the approximate location of $Rotation_1(\triangle ABC)$ followed by $Rotation_2$. Label the image of $\triangle ABC$ as $\triangle A'B'C'$.

b. Describe the sequence of rigid motions that would map $\triangle ABC$ onto $\triangle A'B'C'$.

3. Refer to the figure given below.

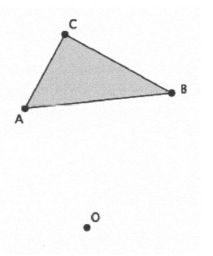

Let R be a rotation of $(-90)°$ around the center O. Let $Rotation_2$ be a rotation of $(-45)°$ around the same center O. Determine the approximate location of $Rotation_1(\triangle ABC)$ followed by $Rotation_2(\triangle ABC)$. Label the image of $\triangle ABC$ as $\triangle A'B'C'$.

Lesson 10: Sequences of Rigid Motions

Classwork

Exercises

1. In the following picture, triangle ABC can be traced onto a transparency and mapped onto triangle $A'B'C'$.
    Which basic rigid motion, or sequence of, would map one triangle onto the other?

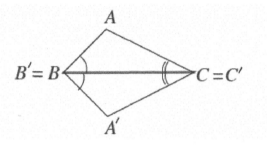

2. In the following picture, triangle ABC can be traced onto a transparency and mapped onto triangle $A'B'C'$.
    Which basic rigid motion, or sequence of, would map one triangle onto the other?

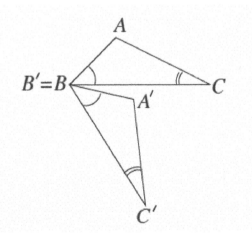

3. In the following picture, triangle ABC can be traced onto a transparency and mapped onto triangle $A'B'C'$.

Which basic rigid motion, or sequence of, would map one triangle onto the other?

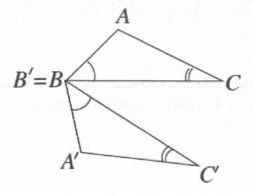

4. In the following picture, we have two pairs of triangles. In each pair, triangle ABC can be traced onto a transparency and mapped onto triangle $A'B'C'$.

Which basic rigid motion, or sequence of, would map one triangle onto the other?

Scenario 1:

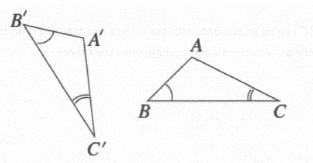

Scenario 2:

5. Let two figures ABC and $A'B'C'$ be given so that the length of curved segment AC equals the length of curved segment $A'C'$, $|\angle B| = |\angle B'| = 80°$, and $|AB| = |A'B'| = 5$. With clarity and precision, describe a sequence of rigid motions that would map figure ABC onto figure $A'B'C'$.

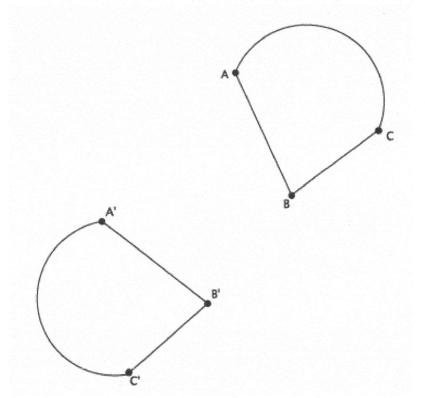

EUREKA
MATH™

Problem Set

1. Let there be the translation along vector $\vec{v}$, let there be the rotation around point A, -90 degrees (clockwise), and let there be the reflection across line L. Let S be the figure as shown below. Show the location of S after performing the following sequence: a translation followed by a rotation followed by a reflection.

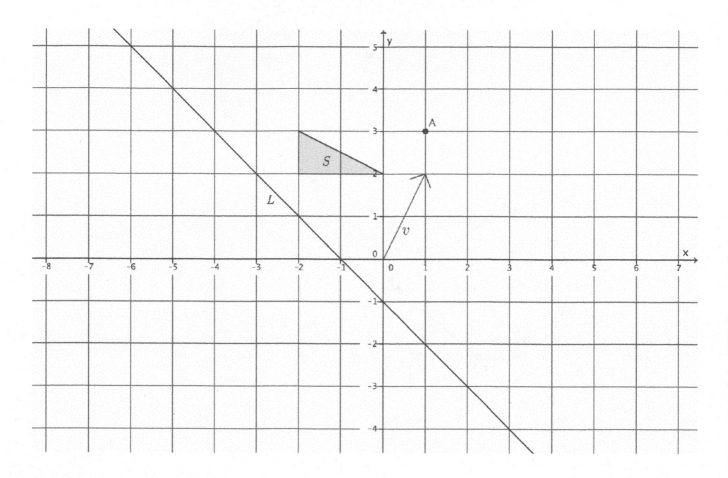

2. Would the location of the image of S in the previous problem be the same if the translation was performed last instead of first; that is, does the sequence, translation followed by a rotation followed by a reflection, equal a rotation followed by a reflection followed by a translation? Explain.

3. Use the same coordinate grid to complete parts (a)–(c).

(coordinate grid with triangle ABC: A at (4, 1), B at (3, 1), C at (3, 4))

a. Reflect triangle ABC across the vertical line, parallel to the y-axis, going through point $(1, 0)$. Label the transformed points A, B, C as A', B', C', respectively.

b. Reflect triangle $A'B'C'$ across the horizontal line, parallel to the x-axis going through point $(0, -1)$. Label the transformed points of A', B', C' as A'', B'', C'', respectively.

c. Is there a single rigid motion that would map triangle ABC to triangle $A''B''C''$?

3. Use the same coordinate grid to complete parts (a)–(c).

a. Reflect triangle A'B'C' across the vertical line, parallel to the y-axis, going through point (1,0). Label the transformed points A', B', C' as A'', B'', C'', respectively.

b. Reflect triangle A''B''C'' across the horizontal line, parallel to the x-axis going through point (0,−1). Label the transformed points of A'', B'', C'' at A''', B''', C''', respectively.

c. Is there a single rigid motion that would map triangle ABC to triangle A'''B'''C'''?

Lesson 11: Definition of Congruence and Some Basic Properties

Classwork

Exercise 1

a.   Describe the sequence of basic rigid motions that shows $S_1 \cong S_2$.

b.   Describe the sequence of basic rigid motions that shows $S_2 \cong S_3$.

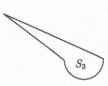

c.   Describe a sequence of basic rigid motions that shows $S_1 \cong S_3$.

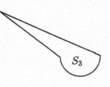

Exercise 2

Perform the sequence of a translation followed by a rotation of Figure XYZ, where T is a translation along a vector $\overrightarrow{AB}$, and R is a rotation of d degrees (you choose d) around a center O. Label the transformed figure $X'Y'Z'$.
Is $XYZ \cong X'Y'Z'$?

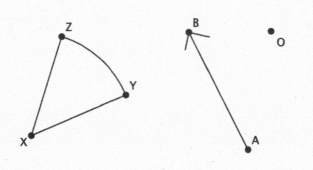

Lesson Summary

Given that sequences enjoy the same basic properties of basic rigid motions, we can state three basic properties of congruences:

(Congruence 1) A congruence maps a line to a line, a ray to a ray, a segment to a segment, and an angle to an angle.

(Congruence 2) A congruence preserves lengths of segments.

(Congruence 3) A congruence preserves measures of angles.

The notation used for congruence is ≅.

Problem Set

1. Given two right triangles with lengths shown below, is there one basic rigid motion that maps one to the other? Explain.

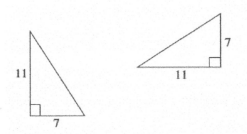

2. Are the two right triangles shown below congruent? If so, describe a congruence that would map one triangle onto the other.

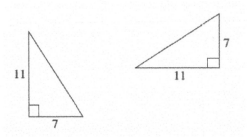

3. Given two rays, $\overrightarrow{OA}$ and $\overrightarrow{O'A'}$:

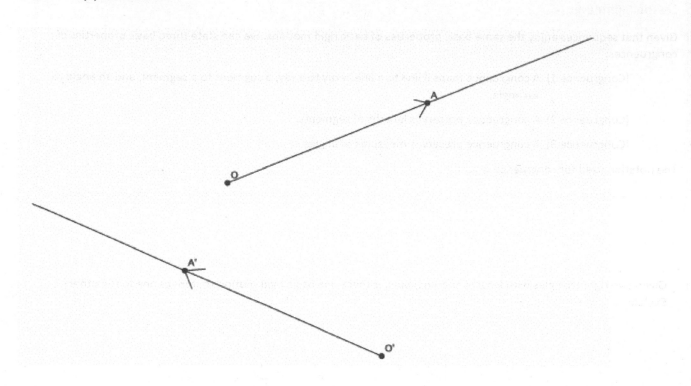

a. Describe a congruence that maps $\overrightarrow{OA}$ to $\overrightarrow{O'A'}$.

b. Describe a congruence that maps $\overrightarrow{O'A'}$ to $\overrightarrow{OA}$.

Lesson 12: Angles Associated with Parallel Lines

Classwork

Exploratory Challenge 1

In the figure below, L_1 is not parallel to L_2, and m is a transversal. Use a protractor to measure angles 1–8. Which, if any, are equal in measure? Explain why. (Use your transparency if needed.)

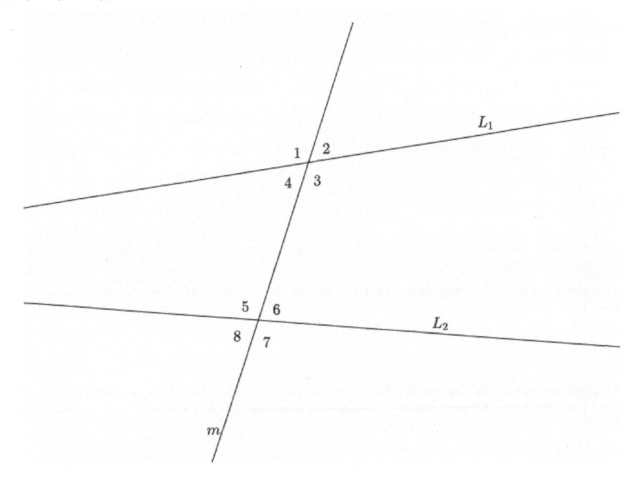

Exploratory Challenge 2

In the figure below, $L_1 \parallel L_2$, and m is a transversal. Use a protractor to measure angles 1–8. List the angles that are equal in measure.

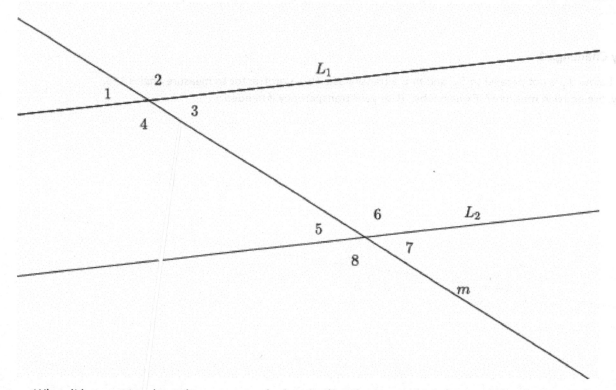

a. What did you notice about the measures of $\angle 1$ and $\angle 5$? Why do you think this is so? (Use your transparency if needed.)

b. What did you notice about the measures of $\angle 3$ and $\angle 7$? Why do you think this is so? (Use your transparency if needed.) Are there any other pairs of angles with this same relationship? If so, list them.

c. What did you notice about the measures of $\angle 4$ and $\angle 6$? Why do you think this is so? (Use your transparency if needed.) Is there another pair of angles with this same relationship?

Lesson Summary

Angles that are on the same side of the transversal in corresponding positions (above each of L_1 and L_2 or below each of L_1 and L_2) are called *corresponding angles*. For example, $\angle 2$ and $\angle 4$ are corresponding angles.

When angles are on opposite sides of the transversal and between (inside) the lines L_1 and L_2, they are called *alternate interior angles*. For example, $\angle 3$ and $\angle 7$ are alternate interior angles.

When angles are on opposite sides of the transversal and outside of the parallel lines (above L_1 and below L_2), they are called *alternate exterior angles*. For example, $\angle 1$ and $\angle 5$ are alternate exterior angles.

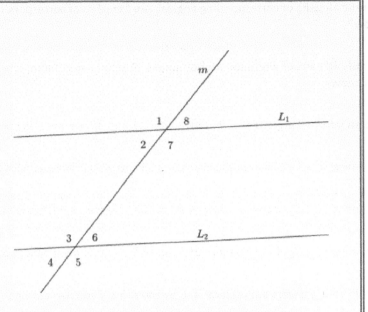

When parallel lines are cut by a transversal, any corresponding angles, any alternate interior angles, and any alternate exterior angles are equal in measure. If the lines are not parallel, then the angles are not equal in measure.

Problem Set

Use the diagram below to do Problems 1–10.

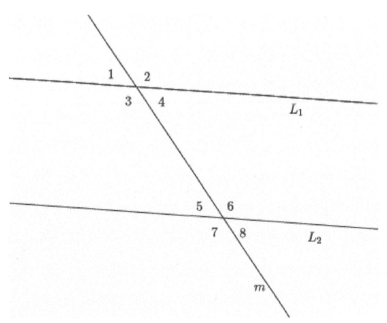

This work is derived from Eureka Math ™ and licensed by Great Minds. ©2015 Great Minds. eureka-math.org

1. Identify all pairs of corresponding angles. Are the pairs of corresponding angles equal in measure? How do you know?

2. Identify all pairs of alternate interior angles. Are the pairs of alternate interior angles equal in measure? How do you know?

3. Use an informal argument to describe why $\angle 1$ and $\angle 8$ are equal in measure if $L_1 \parallel L_2$.

4. Assuming $L_1 \parallel L_2$, if the measure of $\angle 4$ is $73°$, what is the measure of $\angle 8$? How do you know?

5. Assuming $L_1 \parallel L_2$, if the measure of $\angle 3$ is $107°$ degrees, what is the measure of $\angle 6$? How do you know?

6. Assuming $L_1 \parallel L_2$, if the measure of $\angle 2$ is $107°$, what is the measure of $\angle 7$? How do you know?

7. Would your answers to Problems 4–6 be the same if you had not been informed that $L_1 \parallel L_2$? Why or why not?

8. Use an informal argument to describe why $\angle 1$ and $\angle 5$ are equal in measure if $L_1 \parallel L_2$.

9. Use an informal argument to describe why $\angle 4$ and $\angle 5$ are equal in measure if $L_1 \parallel L_2$.

10. Assume that L_1 is not parallel to L_2. Explain why $\angle 3 \neq \angle 7$.

Lesson 13: Angle Sum of a Triangle

Classwork

Concept Development

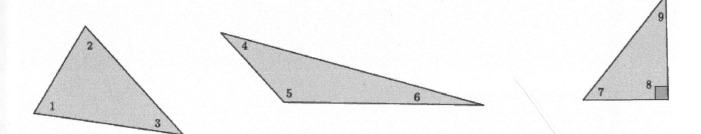

$$m\angle 1 + m\angle 2 + m\angle 3 = m\angle 4 + m\angle 5 + m\angle 6 = m\angle 7 + m\angle 8 + m\angle 9 = 180°$$

Note that the sum of the measures of angles 7 and 9 must equal 90° because of the known right angle in the right triangle.

Exploratory Challenge 1

Let triangle ABC be given. On the ray from B to C, take a point D so that C is between B and D. Through point C, draw a segment parallel to $\overline{AB}$, as shown. Extend the segments AB and CE. Line AC is the transversal that intersects the parallel lines.

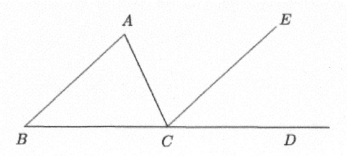

a.   Name the three interior angles of triangle ABC.

b.   Name the straight angle.

c.   What kinds of angles are $\angle ABC$ and $\angle ECD$? What does that mean about their measures?

d.   What kinds of angles are $\angle BAC$ and $\angle ECA$? What does that mean about their measures?

e.   We know that $m\angle BCD = m\angle BCA + m\angle ECA + m\angle ECD = 180°$. Use substitution to show that the measures of the three interior angles of the triangle have a sum of $180°$.

Exploratory Challenge 2

The figure below shows parallel lines L_1 and L_2. Let m and n be transversals that intersec L_1 at points B and C, respectively, and L_2 at point F, as shown. Let A be a point on L_1 to the left of B, D be a point on L_1 to the right of C, G be a point on L_2 to the left of F, and E be a point on L_2 to the right of F.

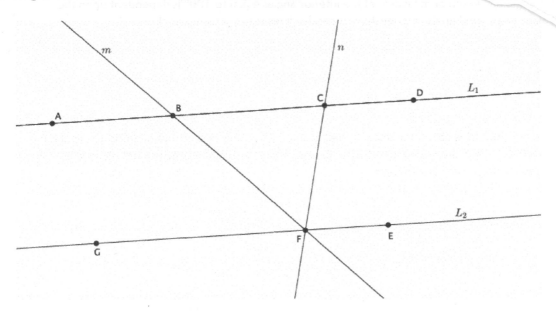

a.   Name the triangle in the figure.

b.   Name a straight angle that will be useful in proving that the sum of the measures of the interior angles of the triangle is 180°.

c.   Write your proof below.

Lesson Summary

All triangles have a sum of measures of the interior angles equal to 180°.

The proof that a triangle has a sum of measures of the interior angles equal to 180° is dependent upon the knowledge of straight angles and angle relationships of parallel lines cut by a transversal.

Problem Set

1. In the diagram below, line AB is parallel to line CD, that is, $L_{AB} \parallel L_{CD}$. The measure of $\angle ABC$ is 28°, and the measure of $\angle EDC$ is 42°. Find the measure of $\angle CED$. Explain why you are correct by presenting an informal argument that uses the angle sum of a triangle.

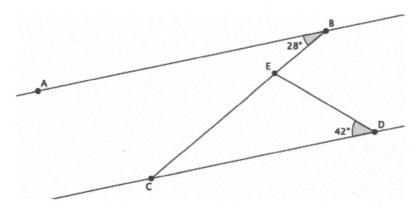

2. In the diagram below, line AB is parallel to line CD, that is, $L_{AB} \parallel L_{CD}$. The measure of $\angle ABE$ is 38°, and the measure of $\angle EDC$ is 16°. Find the measure of $\angle BED$. Explain why you are correct by presenting an informal argument that uses the angle sum of a triangle. (Hint: Find the measure of $\angle CED$ first, and then use that measure to find the measure of $\angle BED$.)

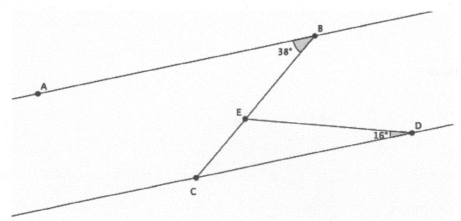

Lesson 13:   Angle Sum of a Triangle

S.76

3. In the diagram below, line AB is parallel to line CD, that is, $L_{AB} \parallel L_{CD}$. The measure of $\angle ABE$ is 56°, and the measure of $\angle EDC$ is 22°. Find the measure of $\angle BED$. Explain why you are correct by presenting an informal argument that uses the angle sum of a triangle. (Hint: Extend the segment BE so that it intersects line CD.)

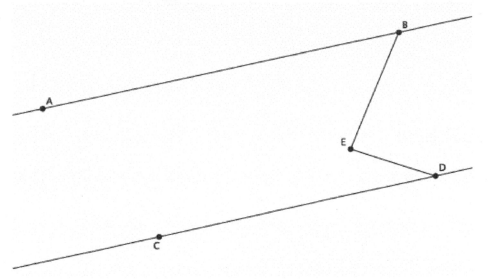

4. What is the measure of $\angle ACB$?

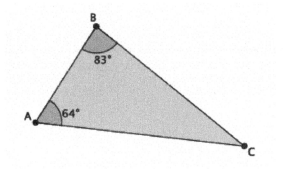

5. What is the measure of $\angle EFD$?

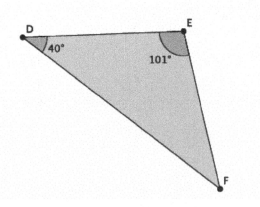

6. What is the measure of ∠HIG?

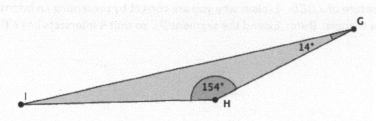

7. What is the measure of ∠ABC?

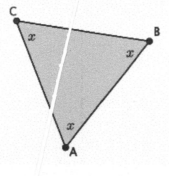

8. Triangle DEF is a right triangle. What is the measure of ∠EFD?

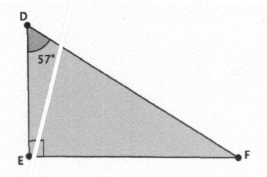

9. In the diagram below, Lines L_1 and L_2 are parallel. Transversals r and s intersect both lines at the points shown below. Determine the measure of $\angle JMK$. Explain how you know you are correct.

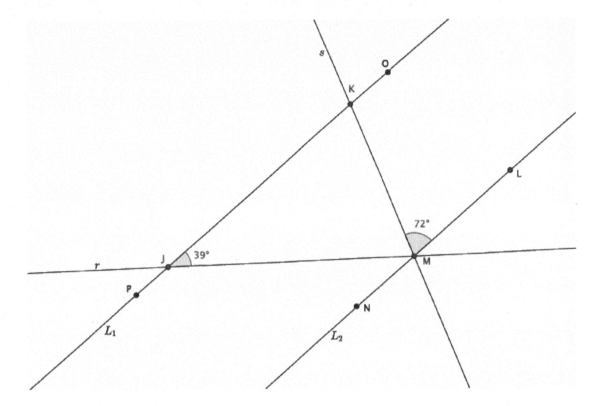

Lesson 14: More on the Angles of a Triangle

Exercises 1–4

Use the diagram below to complete Exercises 1–4.

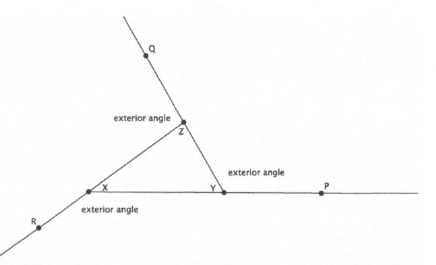

1. Name an exterior angle and the related remote interior angles.

2. Name a second exterior angle and the related remote interior angles.

3. Name a third exterior angle and the related remote interior angles.

4. Show that the measure of an exterior angle is equal to the sum of the measures of the related remote interior angles.

Example 1

Find the measure of angle x.

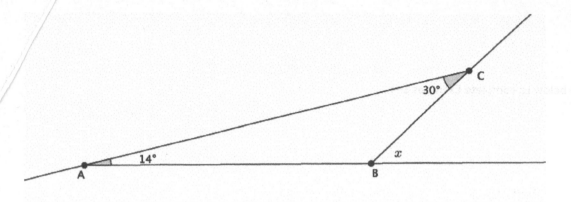

Example 2

Find the measure of angle x.

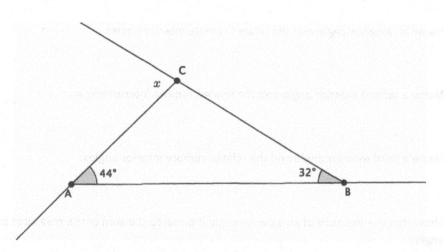

Example 3

Find the measure of angle x.

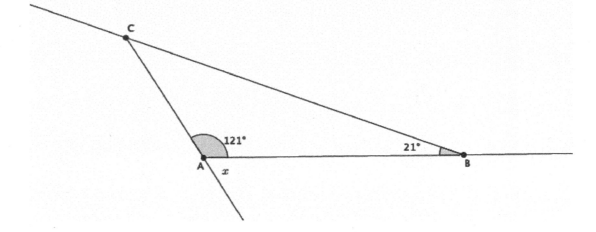

Example 4

Find the measure of angle x.

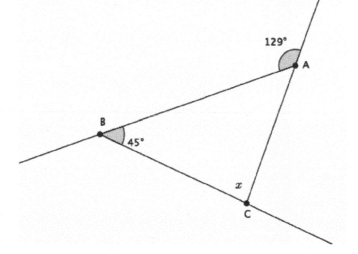

Exercises 5–10

5. Find the measure of angle x. Present an informal argument showing that your answer is correct.

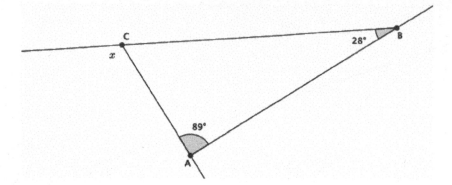

6. Find the measure of angle x. Present an informal argument showing that your answer is correct.

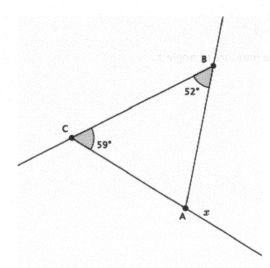

7. Find the measure of angle x. Present an informal argument showing that your answer is correct.

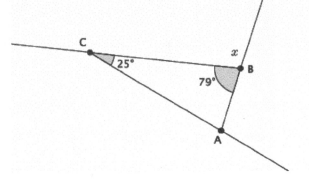

8. Find the measure of angle x. Present an informal argument showing that your answer is correct.

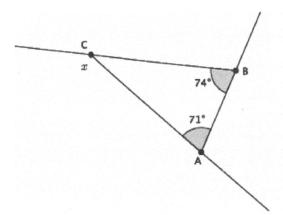

9. Find the measure of angle x. Present an informal argument showing that your answer is correct.

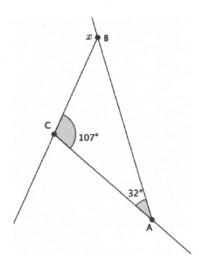

---

10. Find the measure of angle x. Present an informal argument showing that your answer is correct.

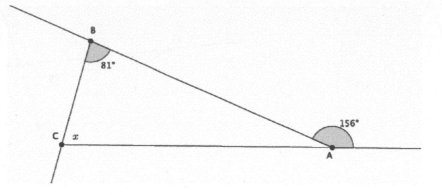

Lesson Summary

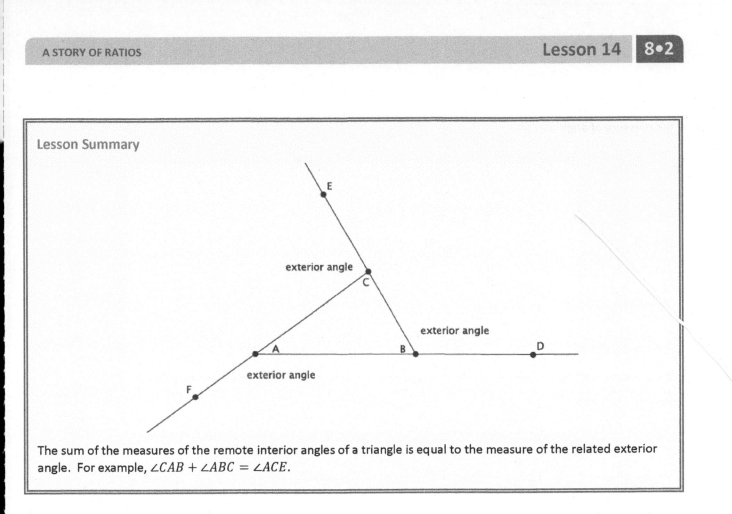

The sum of the measures of the remote interior angles of a triangle is equal to the measure of the related exterior angle. For example, $\angle CAB + \angle ABC = \angle ACE$.

Problem Set

For each of the problems below, use the diagram to find the missing angle measure. Show your work.

1. Find the measure of angle x. Present an informal argument showing that your answer is correct.

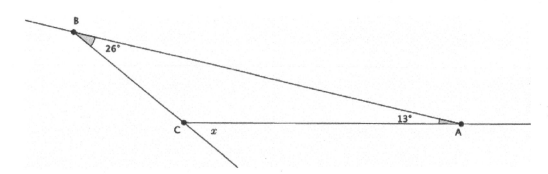

This work is derived from Eureka Math ™ and licensed by Great Minds. ©2015 Great Minds. eureka-math.org

2. Find the measure of angle x.

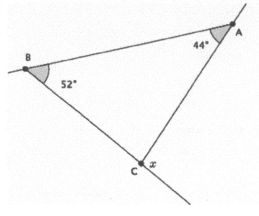

3. Find the measure of angle x. Present an informal argument showing that your answer is correct.

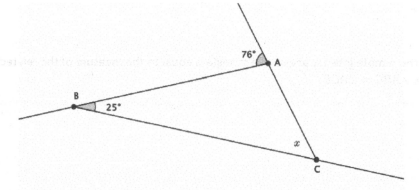

4. Find the measure of angle x.

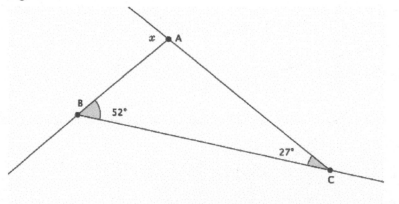

5. Find the measure of angle x.

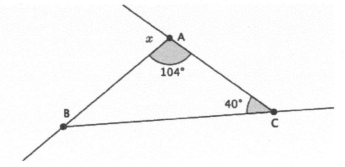

6. Find the measure of angle x.

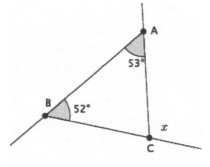

7. Find the measure of angle x.

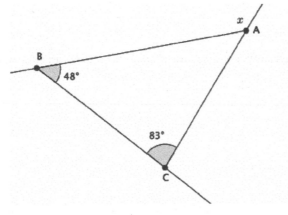

8. Find the measure of angle x.

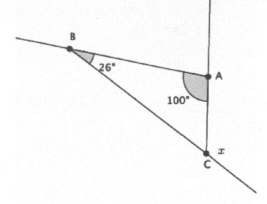

9. Find the measure of angle x.

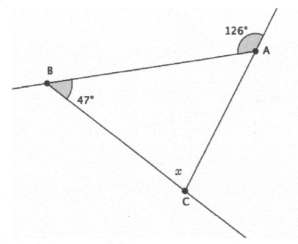

10. Write an equation that would allow you to find the measure of angle x. Present an informal argument showing that your answer is correct.

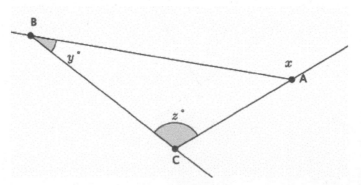

Lesson 15: Informal Proof of the Pythagorean Theorem

Classwork

Example 1

Now that we know what the Pythagorean theorem is, let's practice using it to find the length of a hypotenuse of a right triangle.

Determine the length of the hypotenuse of the right triangle.

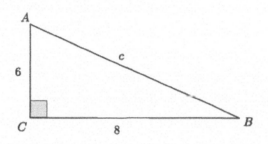

The Pythagorean theorem states that for right triangles $a^2 + b^2 = c^2$, where a and b are the legs, and c is the hypotenuse. Then,

$$a^2 + b^2 = c^2$$
$$6^2 + 8^2 = c^2$$
$$36 + 64 = c^2$$
$$100 = c^2.$$

Since we know that $100 = 10^2$, we can say that the hypotenuse c is 10.

Example 2

Determine the length of the hypotenuse of the right triangle.

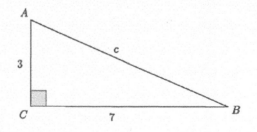

Exercises 1–5

For each of the exercises, determine the length of the hypotenuse of the right triangle shown. Note: Figures are not drawn to scale.

1.

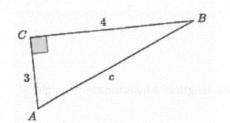

2.

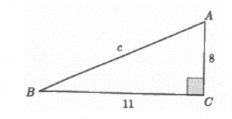

3.

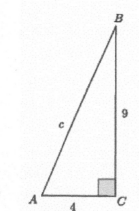

4.

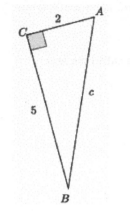

5.

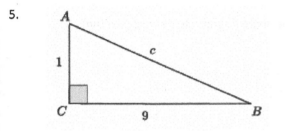

Lesson Summary

Given a right triangle ABC with C being the vertex of the right angle, then the sides $\overline{AC}$ and $\overline{BC}$ are called the *legs* of $\triangle ABC$, and $\overline{AB}$ is called the *hypotenuse* of $\triangle ABC$.

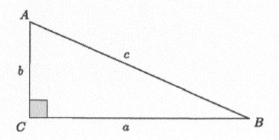

Take note of the fact that side a is opposite the angle A, side b is opposite the angle B, and side c is opposite the angle C.

The Pythagorean theorem states that for any right triangle, $a^2 + b^2 = c^2$.

Problem Set

For each of the problems below, determine the length of the hypotenuse of the right triangle shown. Note: Figures are not drawn to scale.

1.

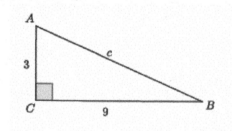

2.

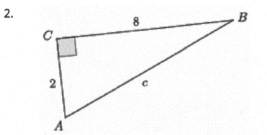

3.

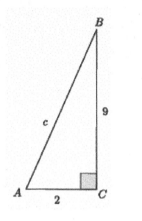

4.

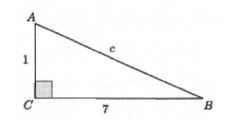

5.

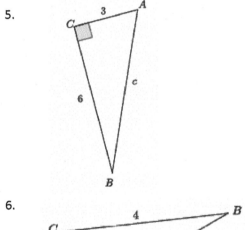

6.

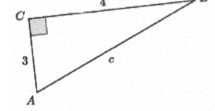

7.

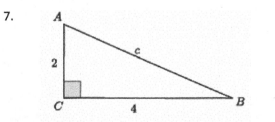

8.

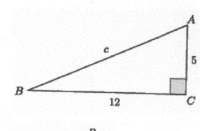

9.

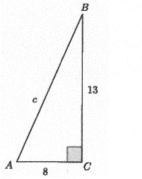

10.

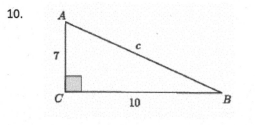

11.

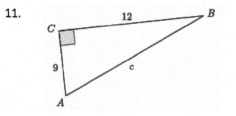

12.

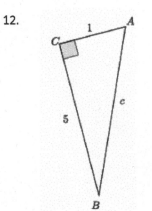

Lesson 16: Applications of the Pythagorean Theorem

Classwork

Example 1

Given a right triangle with a hypotenuse with length 13 units and a leg with length 5 units, as shown, determine the length of the other leg.

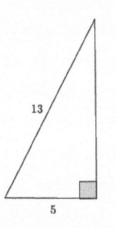

$$5^2 + b^2 = 13^2$$
$$5^2 - 5^2 + b^2 = 13^2 - 5^2$$
$$b^2 = 13^2 - 5^2$$
$$b^2 = 169 - 25$$
$$b^2 = 144$$
$$b = 12$$

The length of the leg is 12 units.

Exercises 1–2

1. Use the Pythagorean theorem to find the missing length of the leg in the right triangle.

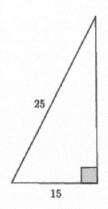

This work is derived from Eureka Math ™ and licensed by Great Minds. ©2015 Great Minds. eureka-math.org

2. You have a 15-foot ladder and need to reach exactly 9 feet up the wall. How far away from the wall should you place the ladder so that you can reach your desired location?

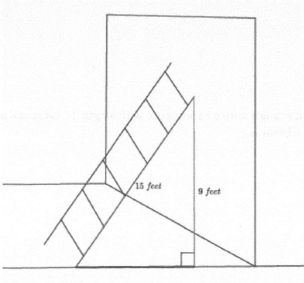

15 feet   9 feet

Exercises 3–6

3. Find the length of the segment AB, if possible.

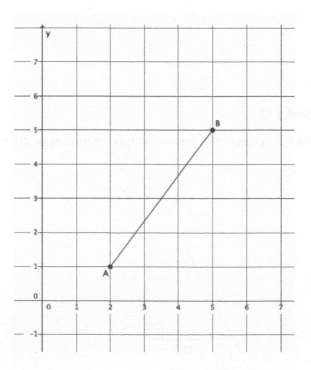

4. Given a rectangle with dimensions 5 cm and 10 cm, as shown, find the length of the diagonal, if possible.

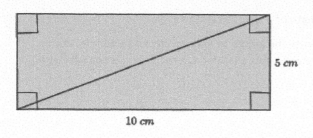

5 cm

10 cm

5. A right triangle has a hypotenuse of length 13 in. and a leg with length 4 in. What is the length of the other leg?

6. Find the length of b in the right triangle below, if possible.

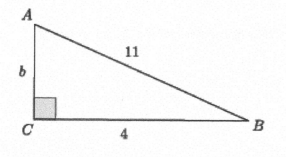

Lesson Summary

The Pythagorean theorem can be used to find the unknown length of a leg of a right triangle.

An application of the Pythagorean theorem allows you to calculate the length of a diagonal of a rectangle, the distance between two points on the coordinate plane, and the height that a ladder can reach as it leans against a wall.

Problem Set

1. Find the length of the segment AB shown below, if possible.

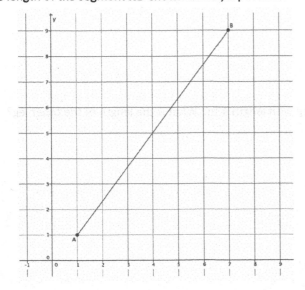

2. A 20-foot ladder is placed 12 feet from the wall, as shown. How high up the wall will the ladder reach?

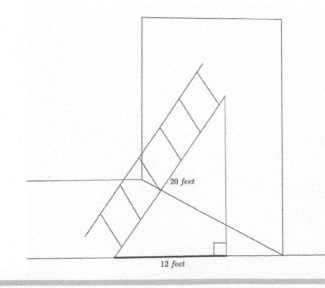

20 feet

12 feet

EUREKA MATH™

Lesson 16:   Applications of the Pythagorean Theorem

3.   A rectangle has dimensions 6 in. by 12 in. What is the length of the diagonal of the rectangle?

Use the Pythagorean theorem to find the missing side lengths for the triangles shown in Problems 4–8.

4.   Determine the length of the missing side, if possible.

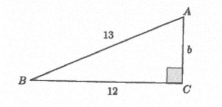

5.   Determine the length of the missing side, if possible.

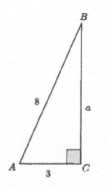

6.   Determine the length of the missing side, if possible.

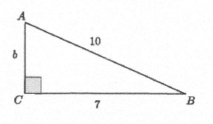

7. Determine the length of the missing side, if possible.

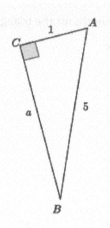

8. Determine the length of the missing side, if possible.

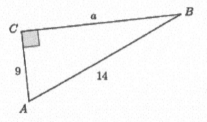